REIT Investing Demystified

Building a Real Estate Portfolio Without a Landlord's Headaches

Joseph Trek

Published by NobleTrex Press

For permissions and other inquiries, write to:

P.O. Box 3132, Framingham, MA 01701, USA

Contents

Introduction

Let's play a game the next time you leave your house.

Walk down the busiest street in your city. Look up. Ignore the coffee shops and the clothing retailers on the ground floor; look at the structures themselves. Look at the massive urgent care center on the corner. Look at the sleek logistics warehouse near the highway that ensures your Amazon packages arrive overnight. Look at the nondescript, windowless concrete fortress that hums with the sound of cooling fans-a data center processing the cloud storage for half the state.

These aren't just piles of brick and steel. They are cash registers.

Every single month, like clockwork, rent checks are cut for these properties. Massive, six-and-seven-figure checks. And unlike the tenant in a duplex who might be late because their car broke down, these tenants are blue-chip corporations, major hospital systems, and government agencies. They sign leases for ten or twenty years. They often pay for their own maintenance, insurance, and taxes.

Now, answer the question that usually stops people in their tracks: **Who owns all of this?**

The cynical answer is "the ultra-wealthy." The faceless institutions. The tycoons with names on library wings. And for a long time, you would have been right. Commercial real estate was a velvet-roped club where the cover charge was ten million dollars and a handshake from a banker.

But that era is over. It has been over since 1960, though most people didn't get the memo.

You can own that data center. You can own the cell tower disguised as a pine tree. You can own the land under the drive-thru pharmacy.

And you can do it with the loose change in your pocket, right from your phone, while waiting for your latte.

Welcome to the empire of the invisible landlord.

The Great Landlord Delusion

If you mention "real estate investing" at a dinner party, the conversation inevitably drifts toward the same exhausted tropes. Someone brings up "flipping" a house. Someone else talks about buying a rental property to generate passive income.

Let's be honest about that second one. We need to shatter a myth before we go any further: **Owning a physical rental property is not passive income.** It is a part-time job that you pay hundreds of thousands of dollars to acquire.

I have friends who are "landlords." I see them answering calls at 11:00 PM because a water heater exploded. I watch them stress over eviction moratoriums, rising lumber costs, and tenants who decided to paint the living room neon orange without asking. They trade their weekends for equity. They confuse "sweat equity" with "investment returns."

There is a strange masochism in personal finance culture that suggests wealth isn't valid unless you suffered for it-unless you sanded the floors yourself.

This book is the antidote to that thinking.

The premise of our time together is simple: You want the check, not the wrench. You want the exposure to the asset class that has created more wealth than perhaps any other in human history-real estate-but you want to strip away the operational friction. You want the profit margin without the phone call.

This is where the Real Estate Investment Trust, or REIT, enters the stage.

The Golden Loophole

To the uninitiated, REITs (pronounced "reets") sound like boring, dusty financial acronyms. In reality, they are one of the most aggressive wealth-compounding vehicles ever created by the U.S. Congress.

Here is the deal the government made: To allow average citizens access to commercial real estate, they created a corporate structure that pays zero corporate income tax. In exchange, these companies must distribute at least 90

Think about what that means. When you buy a share of a regular tech company, the management might hoard the cash, buy a private jet, or waste it on a failed metaverse project. When you buy a share of a REIT, the management is legally compelled to hand the cash over to you.

It turns the stock market upside down. Instead of praying for a stock price to go up based on hype, you are buying into a legally binding stream of cash flow generated by tangible assets.

But this isn't just about collecting a quarterly check. It's about viewing the world through a different lens. Once you understand this market, you stop seeing "buildings." You start seeing economic engines. You realize that the entire modern economy-from the server farms powering Artificial Intelligence to the cold storage facilities keeping our food supply fresh-is a real estate game.

And you can be the house.

Mapping the Territory

We are not going to approach this like a textbook. Textbooks are great for putting you to sleep; they are terrible for making you money. Instead, we are going to approach this like a blueprint for a fortress.

We have to start with the bedrock. Before we buy a single share, we have to look at your personal financial foundation. I know, nobody likes to talk about emergency funds and debt management-it's the vegetables of the finance world. But leverage cuts both ways. You cannot build a long-term real estate portfolio if you are one blown tire away from having to liquidate your account. We will look at the economic engine of real estate itself-how a property actually makes money-so you understand what you are buying.

Then, we enter the candy store. The sheer variety of the REIT universe is staggering. Most people think "real estate" means apartments and office buildings. That's thinking too small. We will tour the sectors that drive the 21st century. We will look at **Industrial REITs** (the backbone of e-commerce), **Infrastructure REITs** (the fiber optics and towers of the digital age), and **Healthcare REITs** (the inevitable demographic wave of an aging population). You will learn why a casino's landlord often makes more reliable money than the casino itself.

But here is where the smart money diverges from the herd. Just because a company is a REIT doesn't mean it's a good investment. There are traps. There are "yield traps"-stocks that offer a seductive
12

To avoid this, I'm going to hand you a new pair of glasses. You cannot read a REIT financial statement the same way you read Apple or Tesla's. Standard metrics like "Earnings Per Share" (EPS) are often meaningless in real estate due to accounting rules regarding depreciation. If you use a P/E ratio to value a REIT, you will get the wrong answer every time. We will learn the language of the pros: **Funds From Operations (FFO)**, **Net Asset Value (NAV)**, and **Cap Rates**. It sounds technical now, but by the time we are done, you'll be able to glance at a balance sheet and know instantly if the dividend is safe.

The Macro and The Micro

Real estate does not exist in a vacuum. It lives and breathes at the mercy of the bond market.

We are going to tackle the monsters under the bed: Interest rates and Inflation. When the Federal Reserve sneezes, real estate catches a cold-or sometimes, a fever. We will demystify the relationship between the cost of debt and property values. You will learn why inflation can actually be a landlord's best friend, provided they have the pricing power to pass it on to tenants.

We will also confront the psychological risks. The stock market is a manic-depressive partner. It will scream at you that your portfolio is worthless one day and priceless the next. Since REITs are traded on public exchanges, they are subject to this volatility. The building didn't shrink, the rent didn't stop coming in, but the ticker symbol turned red. We will build a mental framework to help you distinguish between a broken stock and a broken company.

Finally, we have to put it all together. A pile of bricks is not a house, and a collection of random stocks is not a portfolio. We will discuss construction-how to size your positions, how to diversify across sectors so that a downturn in office space doesn't wipe you out, and most importantly, the tax game.

The government gives with one hand (the REIT structure) and takes with the other (taxes on dividends). We will explore how to locate your assets in the right accounts-IRAs, 401(k)s, or taxable brokerage accounts-to ensure you keep what you earn.

The End of the Landlord's Headache

I wrote this book because I was tired of seeing smart people take the hard road.

I watched friends save for a decade to buy a rental property, only to find themselves painting walls on a Saturday while their "passive" investment drained their bank account for a new roof. I watched others sit in cash, terrified of the stock market, letting inflation eat their retirement alive.

There is a middle path. It is liquid. It is transparent. It is professional.

You are about to learn how to build a global real estate empire from your laptop. You won't need a credit check. You won't need to interview tenants. You won't need to know the difference between PEX and copper piping.

All you need is the discipline to understand the business model, the patience to let compounding do its heavy lifting, and the willingness to think differently about what it means to be an owner.

The toilet is broken in apartment 4B. The roof is leaking at the warehouse. The HVAC system just died at the medical center.

Relax. Someone else is taking care of it. And they are about to send you a check.

Let's get to work.

Chapter 1

The Foundation of Wealth and Real Estate

You wouldn't try to run a marathon in flip-flops, yet many investors rush into the market without the proper footing. This chapter is your warm-up. We will secure your personal financial safety net, uncover the 'magic' of compound interest, and look under the hood of the real estate engine to see exactly how a simple lease turns into a steady paycheck-all before you buy a single share.

1.1 First Things First: The Pre-Investment Safety Check

A REIT portfolio is supposed to feel like *ownership without the midnight plumbing call.* But there's a catch nobody puts in the brochure: markets don't care when your transmission dies. Your investments can be perfectly chosen and still become a problem if your personal finances force you to sell at the wrong time. That's the real enemy here—not volatility itself, but *bad timing imposed on you by real life.* The pre-investment safety check is how you make sure you're never cornered into turning long-term assets into short-term cash at a discount.

Think of it like building a house in hurricane country. You can pick gorgeous countertops later. First you bolt the frame to the foundation. In money terms, that means two unglamorous things: getting rid of expensive, toxic debt and building a liquidity buffer that can absorb a hit. Do those well and REIT investing becomes what it's meant to be: a patient, compounding machine. Skip them and even a sensible portfolio can become a revolving door of panic-selling and "I'll start again next month."

The simple rule Wall Street hates: don't invest at a negative return

Here's a weirdly freeing thought: if you're carrying high-interest debt, you're already investing—just in the bank's portfolio. Credit card debt in particular is the anti-REIT. It compounds too, except it compounds *against* you, it doesn't diversify, and it doesn't send thank-you notes when you make extra payments. It just sits there, quietly turning your future paychecks into someone else's earnings.

Paying down a high APR balance is one of the only places in personal finance where the phrase "guaranteed return" actually belongs. If your credit card charges 20% and you pay off $1,000, you've effectively earned 20% on that $1,000—after taxes, without market risk, without needing a good year in the stock market. To beat that with investing, you'd need an opportunity that reliably nets more than 20% *after* tax and fees, year after year, with no blowups. That doesn't exist in polite society.

The seduction is understandable: REITs pay dividends. You can picture the cash hitting your account. Debt payoff is invisible—nothing arrives, nothing "happens," except your balance shrinks. But your net worth notices. Your stress level notices. And later, when markets inevitably wobble, the difference between "annoying" and "catastrophic" often comes down to whether you still have a minimum payment hanging around your neck like a monthly subscription you can't cancel.

Why high-interest debt is worse than ``risk"

Investing risk gets discussed like it's a character flaw. "I'm not a risk-taker." As if risk is a personality type, like being a morning person or liking olives. In practice, the most dangerous "risk" in early investing is usually *structural*. It's the kind that shows up on your calendar every month.

A credit card bill doesn't care if your employer has a slow quarter. It doesn't care if your kid needs braces or your roof decides to audition for a waterfall. It doesn't care if REIT prices are down 30% for reasons that have nothing to do with your particular holdings. It wants its money on the due date, and if it doesn't get it, the interest meter spins faster and your credit takes a bruise that can cost you in other ways (car insurance, apartment applications, loan rates). That's not "volatility." That's a contract.

So the real comparison isn't "credit card payoff vs. investing returns." It's "a certain 18%–30% headwind vs. the possibility of 8%–10% tailwind with bumps." When you put it that way, the choice stops being a debate and starts being a safety standard.

A debt triage that doesn't require a finance degree

Not all debt is toxic. Some debt is expensive and urgent; other debt is more like a long, low-cost lease on money. The trick is to separate "this is on fire" from "this is a manageable tool."

On fire is typically revolving credit and any loan with a double-digit interest rate. These balances are designed to linger. Minimum payments are engineered to keep you in the chair, feeding the meter, like a casino that also charges you for oxygen.

Manageable tool is usually fixed-rate debt at modest rates, especially when there's an asset behind it (a mortgage, in many cases). That doesn't mean it's always optimal to keep it, but it rarely needs to be treated like an emergency. The interest isn't compounding at a pace that outruns the rest of your life.

A clean approach that works for most people is simple:

- Keep paying minimums on everything so nothing goes delinquent.
- Attack the highest interest rate first (the "avalanche" method).
- If motivation is your biggest obstacle, attack the smallest balance first (the "snowball" method) *only if* you then roll those payments into the next debt instead of celebrating by buying a new problem.

The math prefers the avalanche. Human psychology sometimes prefers the snowball. The best method is the one you will actually execute consistently for months.

If you want a litmus test that cuts through indecision, use this: would you borrow money at your debt's interest rate to invest in REITs? If you wouldn't take out a brand-new 22% loan to buy shares, carrying a 22% balance while investing is the same trade wearing different clothes.

The hidden fee of debt: it makes you a forced seller

Markets don't punish investors for being wrong as often as they punish investors for needing cash at the wrong moment. Debt increases your fixed monthly obligations. Fixed obligations reduce your flexibility. Reduced flexibility turns normal market downturns into personal emergencies.

Picture a scenario: REIT prices drop because interest rates move up and sentiment sours. Your holdings may still be collecting rent through leases and paying dividends, but the market price is down. If your finances are stable, you can wait. If you're juggling high payments, you can't. You sell because you must. That's the opposite of investing—it's pawning.

This is where the "yield" obsession becomes dangerous. A high dividend yield looks like relief when you're squeezed. But reaching for yield while your financial foundation is shaky is like buying the fanciest umbrella while your basement floods. A dividend can be cut. A market price can drop. A credit card bill will not suddenly become more understanding because you owned a high-yield REIT for three months.

The emergency fund: your portfolio's shock absorber

If debt payoff is removing the headwind, an emergency fund is installing a shock absorber. It's the money that keeps you from converting a temporary problem into a permanent setback.

People sometimes treat emergency funds like idle cash that's "not working." That phrase has done more damage than most scams. Cash in an emergency fund isn't there to grow; it's there to *prevent you from doing dumb, expensive things under pressure*. Its job is not to make money. Its job is to keep you from losing money in the worst possible way: selling good assets at bad prices, taking on predatory debt, or skipping necessary repairs that become catastrophic repairs.

An emergency fund is also a sleep aid you don't need a prescription for. The first time you handle an unexpected $1,200 expense without drama, you understand why it matters. The second time, you stop feeling like adulthood is a surprise attack.

How big should it be? Don't count months---count fragility

The standard advice you'll hear is "three to six months of expenses." That's not wrong, but it's incomplete in the way "bring a jacket" is incomplete. A light jacket is fine in October. It's a joke in a blizzard.

A more useful approach is to size your emergency fund based on *fragility*, not vibes. Consider:

- *Income stability:* salaried in a stable role vs. commission-based vs. seasonal vs. gig work.
- *Household structure:* one income vs. two, dependents vs. none.

- *Cost rigidity:* high fixed costs (rent, childcare, medical) vs. flexible spending.
- *Support system:* family help available vs. fully self-supported.

A household with two stable incomes, low fixed expenses, and a strong support network can often live comfortably with a smaller buffer. A household with one income, variable pay, and high fixed expenses is not "being conservative" by building a bigger reserve—it's being realistic.

Here's the key detail that gets missed: the emergency fund should be based on *essential expenses*, not your full lifestyle. Essentials are the bills that keep you housed, fed, insured, and employed: housing, utilities, groceries, basic transportation, minimum debt payments, insurance, required childcare. Essentials are not "everything you paid last month." Essentials are what you'd still pay if you were temporarily operating in survival mode.

A practical target that doesn't feel like climbing Everest

If you're starting from zero, "six months of expenses" can sound like being told to run a marathon when you can't find your sneakers. The fix is to build in layers, where each layer gives you a meaningful win:

- *Layer 1:* $1,000–$2,000 (or one month of essentials, whichever is larger). Enough to handle the most common "life happens" moments.
- *Layer 2:* 3 months of essential expenses. Enough to handle a job transition or a slow season without blowing up your investing plan.
- *Layer 3:* 6+ months for higher fragility households (single income, variable income, specialized job market, dependents).

Notice what's happening psychologically: you stop treating the emergency fund like an enormous monolith and start treating it like a series of safety rails. Every rail you install reduces the chance you'll fall off the investing track.

Where to keep it so it actually works in an emergency

Emergency fund money has two requirements: it must be safe, and it must be accessible. "Accessible" doesn't mean "one thumb tap away in the same app where you buy sneakers." It means you can get it within a day or two without penalties, drama, or market risk.

For most investors, the sweet spot is a high-yield savings account or a money market deposit account. The return won't impress you at parties. That's fine. This money is your financial fire extinguisher. You don't complain that the extinguisher isn't "earning enough."

Keep it separate from your daily spending account if you're prone to "accidental emergencies" (also known as vacations). Separation creates friction, and friction is a feature. It gives you a moment to ask, "Is this truly an emergency, or am I bored?"

The cash-flow system: make investing the default, not the heroic act

Most people don't fail at investing because they're lazy. They fail because investing requires a monthly act of willpower, and willpower has a terrible attendance record. The fix is to stop negotiating with yourself.

A clean cash-flow system is boring in the best way:

- Income lands in a checking account.
- Essentials are paid automatically.
- A set amount flows automatically to the emergency fund until it hits the target.
- Debt payments are automated and prioritized.
- Only after those are handled does money flow to investing.

You're not "being strict." You're removing daily decision-making. You want your financial life to run like plumbing: water goes where it's supposed to go without you standing there every morning with a bucket.

When investing becomes the leftover after random spending, it always loses to random spending. When investing becomes an automatic transfer scheduled right after payday, it stops being a monthly debate and starts being a habit.

The decision tree: where your next dollar should actually go

Right about now, a reasonable person asks: "Okay, but what about *today's* dollar? Where does it go first?" This is where a simple decision tree saves you from the endless mental ping-pong of personal finance.

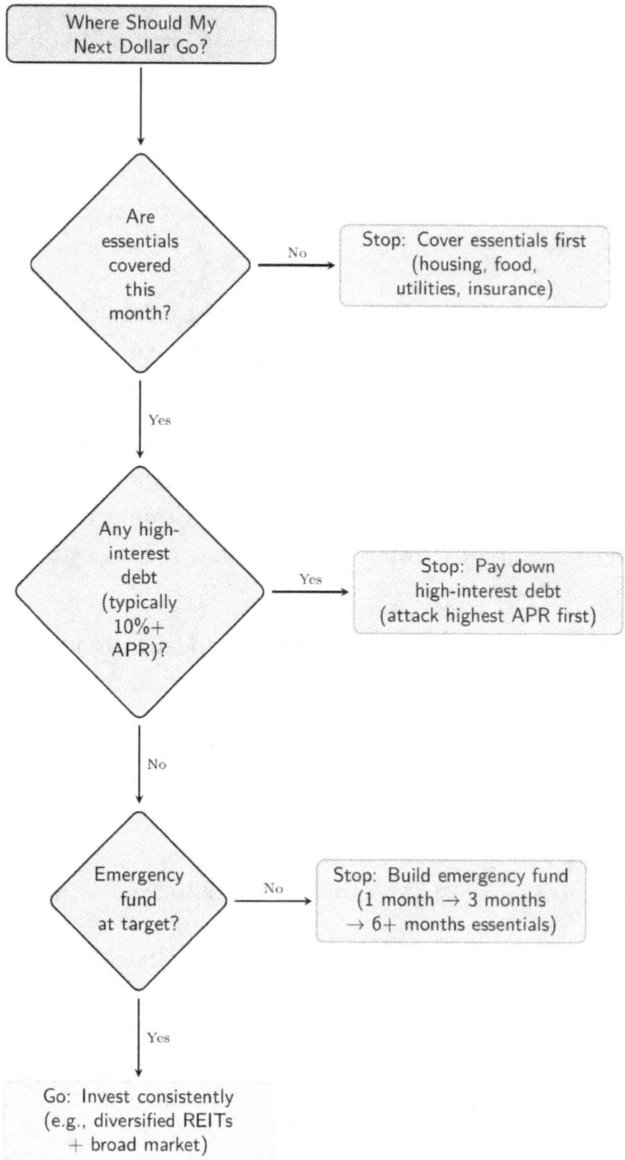

The beauty of a decision tree is that it turns money into a sequence instead of a moral test. You don't have to "feel disciplined." You just follow the flow. When you're tempted to do something flashy, you can literally point to the box that says "Stop" and save yourself from becoming your own worst financial advisor.

The psychological trap: investing feels productive, safety feels boring

There's an emotional reason people skip the safety check. Investing has a scoreboard. You can open an app and see movement. You can join conversations. You can feel like you're building something.

Paying off debt and building an emergency fund feels like sweeping the garage. Useful, necessary, never ending. And yet sweeping the garage is exactly what prevents you from tripping in the dark and breaking your ankle. Boring is often what makes everything else possible.

This is also why "I'll do the emergency fund later" tends to translate into "I'll do it after I've already had a financial emergency." Life has a way of collecting its fees whether you've prepared or not. The emergency fund is simply you choosing the terms.

A quick self-audit: can you handle a bad month without selling?

A pre-investment safety check should end with a question so practical it feels almost rude:

If next month goes sideways, will I be forced to sell investments?

Sideways can mean:

- Your hours get cut.
- You have a medical bill with a high deductible.
- Your car repair shows up with a comma in the price.
- A family member needs help and you're the responsible one.

If the honest answer is "yes," that's not a sign you're failing. It's a sign you've identified the bottleneck. Fix the bottleneck and your investing plan stops being fragile.

The investing plan that survives real life

Once toxic debt is shrinking and the emergency fund is rising, something subtle but powerful happens: you begin investing from a place of strength instead of desperation. You're no longer chasing dividends because you need them to plug a hole. You're choosing investments because they fit your goals.

That's the point where REITs start to behave the way long-term investments are supposed to behave. A price drop becomes an inconvenience, not a threat. A dividend becomes a component of total return, not a lifeline. Volatility becomes background weather, not an existential crisis.

And that's when the real advantage of public REITs shows up: you can own diversified real estate while keeping your personal finances liquid and resilient. The goal isn't to avoid risk entirely. The goal is to avoid *avoidable* risk—the kind created by high-interest debt, thin cash reserves, and a system that forces you to make decisions under pressure.

A final word on impatience (because everyone has it)

The urge to "get started now" is not a flaw. It's a sign you're awake. Just don't confuse motion with progress. If your financial foundation is shaky, piling investments on top isn't building wealth; it's stacking boxes on a wobbly table.

Handle the expensive debt. Build the buffer. Automate the flow. Then invest like someone who can afford to be right over time—which is, quietly, the only kind of investing that tends to work.

1.2 The Eighth Wonder: Compounding and the Long Game

Compounding is the closest thing investing has to a cheat code, except it doesn't feel like cheating while it's happening. It feels like nothing. For years it can look like you're pushing a heavy object up a shallow hill: the numbers move, but they don't *impress*. Then, almost rudely, time flips the script. The same habits that once seemed modest start producing results that look like talent.

The reason compounding gets treated like folklore is that it's hard to *feel* exponential growth in real time. Human intuition is linear. We're good at "add a little each month." We're bad at "growth that accelerates because the thing that grew last year gets to grow again." That second part—growth on prior growth—is where wealth separates from wages.

A REIT portfolio is a particularly clear stage for this drama because the cash flows are visible. Rents become income. Income becomes dividends. Dividends can be spent, or they can be fed back into the machine to buy more shares that earn more dividends that buy more shares. If you reinvest, you're not just holding real estate. You're building a self-replicating income stream with tiny, compounding bricks.

The moment money starts earning money (and then starts earning *more* money)

Most people understand interest. You have $100 and it earns 5%, so you make $5. That's simple interest: the original principal earns a return.

Compounding is what happens when you don't remove the $5. You leave it there. Now you have $105, and next year the 5% applies to $105. Your money earns money, and then it earns money on the money it earned. If you like concrete numbers, here's the cleanest example:

- Start: \$100
- After 1 year at 5%: \$100 × 1.05 = \$105
- After 2 years at 5%: \$105 × 1.05 = \$110.25

That extra \$0.25 in year two isn't magic. It's the first visible footprint of compounding. It's small on day one. That's the trap. Compounding is always shy at the beginning because the base it's multiplying is still small.

The general formula looks scarier than it is:

$$FV = PV \times (1 + r)^n$$

where PV is what you start with (present value), r is the annual return, and n is the number of years. The whole story is in that exponent. The return doesn't just *add* each year; it *stacks*.

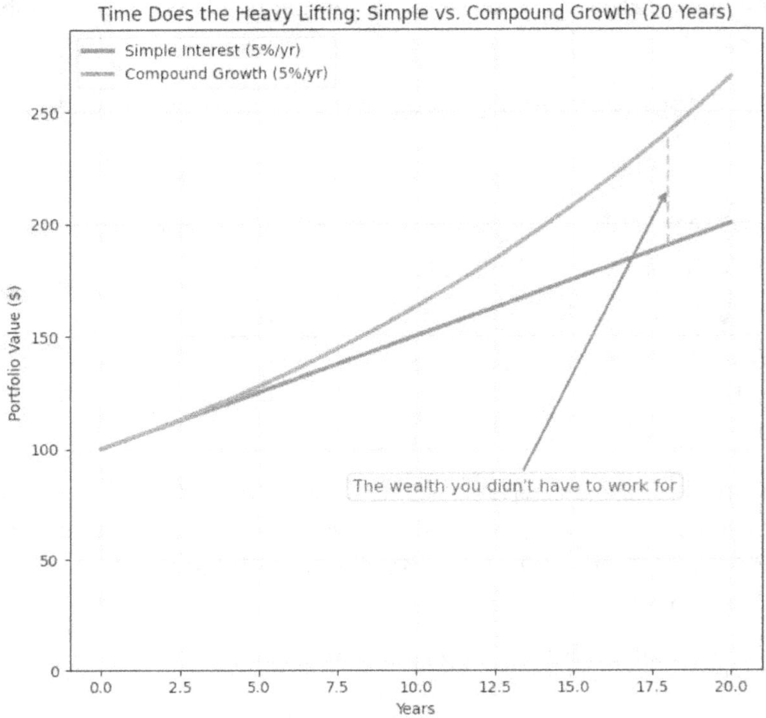

Illustration only (hypothetical constant return). Real markets fluctuate.

That widening gap in the chart isn't a motivational poster. It's a mathematical consequence. In the early years, the two lines behave like distant cousins who share the same last name but don't really talk. In the later years, compounding starts acting like a different species. The curve isn't "better because it tried harder." It's better because it had *time to multiply*.

Why reinvesting dividends is the quiet superpower of REITs

REITs are often introduced through their dividend yields because dividends are tangible. You can point to them. You can count them. You can build a spreadsheet that makes you feel like you have a small financial staff.

The deeper point is what dividends *enable*: reinvestment with discipline. If you reinvest dividends (often abbreviated as DRIP, for Dividend Reinvestment Plan), you're doing two things at once:

- keeping your ownership stake growing without adding fresh cash, and
- increasing future dividend dollars because you own more shares.

This is why a REIT portfolio can become emotionally addictive in a healthy way. Your dividends become a scoreboard for patience. In year one, reinvestment buys a few crumbs of extra shares. In year ten, those crumbs have had time to grow into slices. The temptation is to judge reinvestment by the first few months. That's like judging an oak tree by its first week.

A useful mental image is a snowball rolling downhill, except without the cliché. Imagine a small business that uses its profits to buy new equipment that increases production that increases profits that buys more equipment. Reinvested dividends work the same way, but you don't have to manage employees or negotiate with suppliers. Your role is to keep the loop intact.

The long game is mostly a behavior problem, not an intelligence problem

If compounding is so powerful, why don't more people benefit from it?

Because compounding makes a brutal demand: *it requires you to look wrong for a while.* Not wrong as in "losing money forever." Wrong as in "nothing exciting is happening," or "my friend doubled their money in a meme stock," or "my account went down this year and now I feel personally insulted."

The long game has an awkward social cost. It's boring to talk about consistent contributions. It's fun to talk about the time you nailed a trade. Nobody tells stories at dinner parties about their well-diversified portfolio's steady reinvestment policy. And yet the dinner party stories are rarely the ones that build wealth.

This is where REIT investors can accidentally stumble into good behavior. Many REIT holders come for income. Income investors tend to pay attention to cash flow, not just price. That focus can help you hold through volatility, because a dividend check feels like a reason to stay seated. It's not foolproof, but it's a sturdier anchor than staring at a red number on a screen.

Risk: not the danger of losing money, but the price of admission

People use the word "risk" like it means "the probability of ruin." In everyday investing, especially with diversified public assets, risk usually shows up as *volatility*: the market price bouncing around while the underlying businesses keep operating.

That distinction matters because volatility feels like danger even when it's not. A REIT's share price can drop quickly because interest rates rise, or because the market is in a bad mood, or because investors suddenly decide they prefer tech stocks again. The properties didn't vanish. The tenants didn't all move out overnight. But your screen turns red, and your nervous system treats it like a fire alarm.

The "price of admission" framing is useful because it's honest. You don't get long-term returns without enduring periods where the market tests your patience. Volatility isn't proof that you made a mistake. It's often proof that you bought a real asset that trades in a real market with real humans—humans who panic, extrapolate, overreact, and occasionally behave like caffeinated squirrels.

The goal isn't to eliminate volatility. The goal is to set your financial life up so volatility can't *force your hand*. Once you've done the safety work, volatility becomes something you can observe instead of something that controls you.

A quick shortcut: the Rule of 72 (and why it's both useful and dangerous)

There's a popular back-of-the-napkin tool called the Rule of 72: divide 72 by your annual return and you get a rough estimate of how many years it takes your money to double.

At 8% per year, $72 \div 8 \approx 9$. That suggests doubling about every nine years. At 6%, $72 \div 6 = 12$. That suggests doubling about every twelve years.

This shortcut is useful for building intuition. It turns "returns" into "time," which is what compounding actually eats. It also exposes the brutal cost of shaving a couple percentage points off your long-term return with fees, bad timing, or chronic panic.

It's also dangerous if you use it as a promise. Returns aren't metronomes. Markets don't pay you 8% every year like a polite subscription service. Some years are great, some are ugly, and occasionally you get a year that makes you question every decision you've ever made. The Rule of 72 is a *compass*, not a calendar.

Total return: the grown-up way to think about dividends

Dividend yield gets the attention because it's easy to quote. "This REIT yields 6%." Clean. Satisfying. But yield alone is like judging a restaurant by the bread basket. Nice, not the whole meal.

The more complete lens is *total return*: the combination of (1) price change and (2) dividends (and whether you reinvest them). A REIT can have a high yield and a poor total return if the share price drifts down over time or the dividend gets cut. Another REIT can have a lower yield but excellent total return if the underlying cash flows grow and the market rewards that growth.

This matters because the market is perfectly capable of luring investors into what looks like income but is actually a slow leak. A very high yield can be a warning label, not a gift. Sometimes it's high because the market expects trouble. Sometimes it's high because the payout is too aggressive for the cash flow. Sometimes it's high because the business is genuinely strong and temporarily out of favor—but you don't get to assume that. You have to investigate.

Total return thinking keeps you honest. It forces the question: "Am I getting paid because the business is healthy, or because the price has fallen and the market is nervous?"

The emotional math of a drawdown

The market has a nasty habit of teaching the same lesson repeatedly: losses feel larger than gains. A 20% drop hurts in a way a 20% gain doesn't heal. That's not weakness; it's biology.

But the math of recovery is non-negotiable. If something falls 50%, it needs to rise 100% to get back to even. That asymmetry is why avoiding catastrophic mistakes matters so much more than optimizing for brag-worthy wins. It's also why leverage and concentration can turn "normal volatility" into "permanent setback."

REITs can be volatile, especially around interest rate shifts. That's not inherently bad. It's a reminder to respect position sizing and diversification, and to treat any single holding as something that can misbehave for longer than you'd like.

The long game is less about being fearless and more about having a plan that doesn't require fearlessness. You want a plan that works even when you're tired, busy, and slightly pessimistic—which is most of adulthood.

Consistency beats cleverness (and yes, that's annoying)

A lot of investing advice is secretly entertainment. It sells the idea that the key to wealth is a special insight, a perfect entry point, or a clever sector rotation. Those things can matter at the margins, but they are not the engine.

The engine is embarrassingly plain:

- contribute consistently,
- reinvest when it makes sense for your goals,
- diversify to avoid a single bad bet ruining your timeline, and
- give the math enough time to stop being shy.

The boring parts are the parts that scale. You can do them in a busy year. You can do them when the news is chaotic. You can do them without needing to be the smartest person in the room.

If you want a concrete example of how consistency wins, think about the investor who buys in perfect timing once every few years versus the investor who buys regularly through good markets and bad. The regular investor ends up purchasing shares at a range of prices. They don't need to predict the future; they just need to keep showing up. In the long run, showing up is a competitive advantage because so many people can't.

How compounding changes what "patience" really means

Patience in investing gets framed like saintly restraint: resisting temptation, mastering your emotions, staring serenely into the abyss. In real life, patience is more practical. It's the willingness to keep doing a sensible thing while it looks unimpressive.

Compounding rewards that specific kind of patience because it doesn't pay you proportionally. It pays you *disproportionately*, later. The early years are the down payment. The later years are the house.

This is why starting earlier matters so much, even if the early contributions are small. A smaller amount invested earlier can beat a larger amount invested later simply because the earlier dollars get more years of multiplication. Time doesn't just increase returns; it changes their *shape*.

If that feels unfair, good. It is unfair—in the same way gravity is unfair. You don't negotiate with it. You position yourself to benefit from it.

A realistic expectation for REIT investors: smooth income, bumpy prices

REIT investors often get surprised by a specific combination: dividends that look relatively steady and prices that can swing around like they've had too much espresso. That combination is not a contradiction; it's a feature of publicly traded real estate.

Properties tend to generate rent with some stability because leases don't reprice every day. Public markets, on the other hand, *reprice every second*. They respond to interest rates, macro fears, sector narratives, and flows in and out of funds. So you can have a REIT whose properties are operating fine while the stock price is having a public meltdown.

Understanding this helps you interpret what you see:

- Dividend stability usually reflects cash flow stability (not guaranteed, but often anchored by leases).
- Price volatility often reflects changing discount rates and sentiment (which can move faster than fundamentals).

This is where the long game becomes an advantage rather than a slogan. If your timeline is measured in years, not weeks, price volatility is more like a sale than a verdict. Not always—sometimes price drops are warnings. But volatility alone is not evidence of failure. It's evidence that you're holding something real in a market that is sometimes irrational.

The best mental shift: from ``What will this do next?" to ``What will this do *over time*?"

Most investing anxiety comes from asking a question nobody can answer: "What will happen next?" Even professionals with expensive models and impressive suits are guessing more than they'd like to admit. Short-term prediction is the part of investing where confidence is cheap and accuracy is rare.

A better question is: "If I keep owning this and reinvesting sensibly, what is the likely outcome over a decade?" That question favors business quality, valuation discipline, and staying power. It also aligns with how compounding actually works.

With REITs, the "over time" lens naturally leads to the fundamentals that matter most: the durability of rents, the ability to raise rents (or grow through development and acquisitions), sensible balance sheets, and dividend policies that don't depend on miracles. Price matters, but time changes the meaning of price. Over long periods, returns tend to be pulled toward underlying cash flows like a magnet.

Compounding doesn't remove uncertainty. It just makes the payoff of sensible behavior large enough that you don't need to be perfect. You need to be consistent. You need to survive. You need to give time a chance to do what time does.

And when you do, the "eighth wonder" stops being a quote and becomes something more satisfying: a quiet, accumulating advantage that's hard to see in year two and almost impossible to ignore in year twenty.

1.3 The Landlord's Ledger: How Properties Actually Make Money

A rent check is not profit. It's *"raw material"*.

That sounds obvious until you watch how many otherwise smart investors talk about real estate like it's a vending machine: insert property, receive monthly cash. Real property is closer to a small business with a moody building attached. Money comes in at the top, sure—but it gets skimmed, siphoned, and occasionally mugged on the way down. The only way to understand a REIT (or a duplex, or a strip center) is to follow the dollars as they fall through the building like water through a set of pipes.

Do that once and a lot of *"mysteries"* stop being mysterious. Why does one landlord brag about a $2,500 rent and another landlord shrug at $2,500 rent? Why can two apartment buildings on the same street have wildly different values? Why do REIT earnings look weird compared to normal companies? The answers live in the ledger—not the ticker symbol.

Gross rent: the number that gets all the attention (and deserves suspicion)

Gross rent is the headline. It's the *"this place brings in $X per month"* number that makes real estate sound like an ATM with better wallpaper. It's also the number most likely to mislead you, because it ignores the simple fact that buildings are hungry.

A property doesn't just *collect* rent; it *consumes* money to keep collecting rent. Roofs age. Parking lots crack. Elevators develop expensive opinions. Tenants move out and new tenants demand fresh paint and "a more modern vibe," as if drywall has feelings. If you treat gross rent like profit, you'll feel rich right up until the first real repair invoice arrives and your *"cash flow"* discovers gravity.

Gross rent is still useful. It tells you the property's earning potential. But it's the top of the waterfall. You don't measure a river by staring at the clouds.

The leaks in the bucket: where rent goes to disappear

Every income property has leaks, and the leaks aren't a sign the property is broken. They're the cost of doing business. Ignore them and you don't get a cleaner analysis; you get fiction.

The big leaks tend to come in a few familiar shapes:

Taxes. The local government is a partner you didn't invite and can't fire. Property taxes can rise even when your rent doesn't, and they don't negotiate the way a contractor sometimes will. They also vary drastically by location, which is one reason "the same" building can behave very differently in different cities.

Maintenance and repairs. Some of this is predictable: landscaping, cleaning, routine servicing. Some is a jump scare: water intrusion, HVAC failure, a plumbing issue that turns into an archeological dig. The key point is that maintenance isn't optional if you want the property to keep producing rent. Tenants are customers, and customers don't keep paying for a product that deteriorates.

Insurance. It's the bill you're happiest to pay when you need it and least happy to pay when you don't. Insurance costs can surge after disasters, and in some regions it can become a major driver of property economics.

Utilities and services. Depending on the property type and lease structure, the owner may pay for water, trash, common-area electric, security, snow removal, pest control, and more. A building with a lobby is a building with bills.

Management. If you manage yourself, you pay in time and stress. If you hire management, you pay in fees. Either way, the work exists. The building doesn't run on vibes.

And then there's the leak that people recognize but consistently underestimate:

Vacancy. Vacancy is not just *"no rent."* Vacancy is often *negative cash flow*, because you may still be paying taxes, insurance, utilities, and sometimes tenant improvement costs to attract the next occupant. Vacancy is why landlords care so much about tenant quality. It's not snobbery. It's math.

The metric that makes real estate make sense: NOI

This is where the ledger gets useful. The industry's workhorse metric is *Net Operating Income*, usually written as NOI.

NOI is the money a property produces after paying the necessary operating expenses required to run it, but before paying for financing and certain other owner-level costs. In plain English: it's the property's profit from operations. Not perfect. Not final. But close enough to compare one property to another without getting distracted by how each owner chooses to finance it.

A clean way to say it:

$$\text{NOI} = \text{Property Revenue} - \text{Operating Expenses}$$

Property revenue usually includes rent and may include things like parking fees, laundry income, late fees, and reimbursements (depending on the lease setup). Operating expenses include the ongoing costs to keep the property functioning and rentable: taxes, insurance, repairs and maintenance, management, and similar items.

NOI deliberately ignores a few huge categories:

- *Debt service* (principal and interest payments) because NOI is meant to be comparable across different financing choices.
- *Depreciation* because it's an accounting expense, not a cash expense.
- *Income taxes at the owner level* because different owners have different tax situations.
- Often *capital expenditures* (big-ticket replacements like roofs) depending on the context; many analysts treat these separately as *CapEx*.

That list is not a loophole; it's a purpose. NOI is trying to answer a specific question: *How well does this building earn money as a building?*

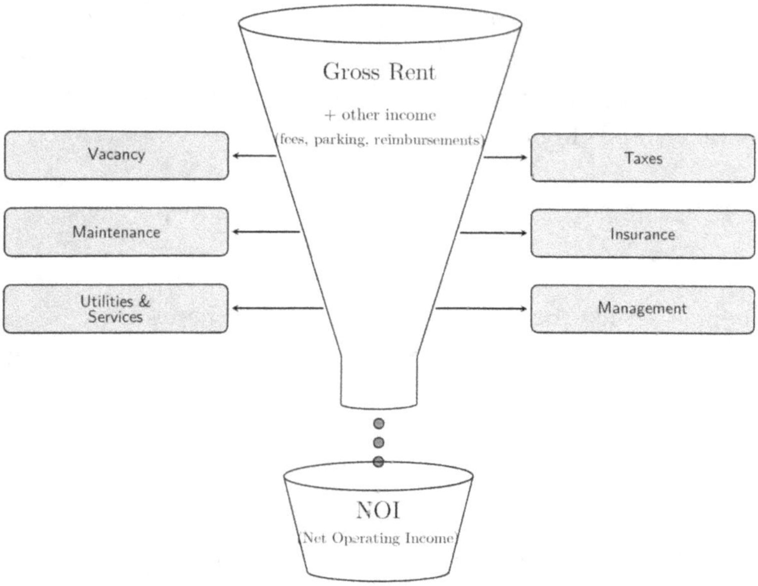

That funnel is the entire real estate business model in one picture. A property is a system for turning *gross rent* into *NOI* while constantly fighting the leaks. Great operators don't magically eliminate leaks; they control them, anticipate them, and price their leases and renovations knowing the leaks are coming.

A tiny story that explains 90% of landlord math

Imagine a small apartment building that collects $20,000 per month in gross rent. Sounds great. Now do the unromantic subtraction.

Maybe $3,000 goes to property taxes, $1,200 to insurance, $2,500 to maintenance and repairs averaged over the year, $1,000 to management, $800 to utilities and common-area costs, and you plan for

$1,500 in vacancy and turnover costs on average. Suddenly your $20,000 becomes roughly $10,000 in NOI.

That's not pessimism. That's adulthood.

And notice what happened: a property with "$20k in rent" and a property with "$10k in NOI" are the same property. The difference is whether you're looking at the top of the funnel or the bottom. Real estate investors live at the bottom.

Lease structures: who pays which bills changes everything

If you've ever wondered why commercial real estate people talk about leases like they're arguing over scripture, here's why: the lease determines who pays which expenses. Two buildings with the same tenants and the same rent can produce very different NOI depending on what the landlord is responsible for.

In some commercial leases, tenants reimburse the landlord for property taxes, insurance, and maintenance (often called *"triple-net"* or NNN, because it typically refers to three major categories). In others, the landlord shoulders more of the expense burden and charges higher rent in exchange. In apartments, the owner often pays many costs directly, though tenants may pay some utilities depending on the setup.

This is why "rent growth" and "NOI growth" are not always the same. If rents rise but expenses rise faster, NOI can stagnate. If rents rise and the landlord has strong expense controls (or the lease pushes expenses to the tenant), NOI can surge. The lease is the rulebook for how the funnel behaves.

Capex: the expenses that don't show up until they do

Operating expenses are the daily diet of a property. Capital expenditures—CapEx for short—are the occasional, larger bites: roofs, paving, major HVAC replacements, elevators, structural work, large-scale unit renovations.

CapEx is where naive cash-flow projections go to die, because it's easy to ignore and hard to argue with. Nobody wants to budget for a roof replacement when the roof is still, technically, keeping most of the rain outside. But buildings are patient. They wait until your timing is inconvenient.

Many property-level analyses treat NOI as separate from CapEx because NOI is meant to capture ongoing operations, while CapEx is lumpy and tied to long-term asset preservation or upgrades. Still, an investor who ignores CapEx is like a driver who ignores oil changes because the engine hasn't complained yet. The complaint arrives later, louder, and much more expensive.

For REIT analysis, this is one reason you'll see metrics beyond NOI that try to approximate "cash available after keeping the properties competitive." You don't need to memorize the alphabet soup yet, but you do need to respect the concept: the bottom of the funnel isn't always the same as the cash you can safely pull out.

Why GAAP earnings make REITs look strange (and why that's not a conspiracy)

Here's where the landlord ledger collides with stock market accounting.

Most companies buy machines that wear out. Accounting reflects that wear through depreciation—a non-cash expense that spreads the cost of an asset over its useful life. Real estate is weird because buildings *also* wear out, but the economics don't always match the accounting story. A well-located property can become more valuable over time even while accounting depreciation marches downward like a metronome.

That accounting mismatch can make a healthy REIT look like it has low earnings, or even negative earnings, despite producing strong property cash flows. To help investors compare operating performance, the REIT industry widely uses a supplemental (non-GAAP) measure called *Funds From Operations*, or FFO.

FFO is not a magical *"true earnings"* number. It's a standardized adjustment meant to reduce distortions caused by real estate depreciation and certain gains or losses from property sales. In spirit, it's trying to answer: *What did the REIT's properties generate from operations, before the accounting treatment of depreciation clouds the picture?*

A crucial warning, because it keeps people honest: non-GAAP measures like FFO are meant to *complement* GAAP statements, not replace them. If someone uses FFO as an excuse to stop reading the financials, they've missed the point. Supplemental metrics are headlights, not autopilot.

Why NOI is the bridge between "real estate" and "investing"

NOI is more than a line item. It's the language that lets you translate a physical building into an investable cash-flow stream.

Once you have NOI, you can:

- compare properties across markets and property types,
- evaluate operational improvements (raise rents, reduce expenses, cut vacancy),
- and connect property performance to valuation tools used throughout commercial real estate.

It's also the anchor that keeps you from being hypnotized by superficial numbers. *"This REIT yields 7%"* is interesting. *"This REIT's properties are growing NOI because rents are rising faster than expenses, and vacancy is controlled"* is *understanding*. One is a surface statistic. The other is the engine.

Tenant as customer: the most underrated sentence in real estate

It's fashionable to talk about tenants like they're obstacles. That attitude is a great way to end up with vacancy and legal fees.

In economic terms, tenants are the customer. Their rent is revenue. Their renewal is retention. Their move-out is churn. A landlord's job is to offer a product (space) that customers will pay for, and to maintain it so customers keep paying.

This is why "nice" properties often make more money than "cheap" ones in ways that aren't obvious at first glance. A well-maintained building can attract better tenants, reduce turnover, and lower the expensive chaos of constant make-readies. It can also support rent increases because the product is genuinely better. The expenses are higher, yes—but the leaks can be smaller where it counts most: vacancy and turnover.

For REITs, the same idea plays out at scale. A well-run REIT is not just a pile of buildings. It's a customer-retention machine with a balance sheet.

A landlord's ledger mindset makes you a better REIT investor

Public markets love shortcuts: dividend yield, price charts, hot takes. The landlord's ledger insists on reality. It asks:

- Where does the money actually come from?
- What are the recurring leaks?
- How stable is occupancy?
- Can rents rise without losing tenants?
- Are expenses controlled or exploding?

Ask those questions and you stop treating REITs like lottery tickets with quarterly payouts. You start seeing them as operating businesses whose value comes from a very old, very stubborn idea: someone pays to use space, and the owner keeps enough of that payment after expenses to make ownership worthwhile.

Once you can follow the funnel from gross rent down to NOI, you can read a REIT's story without needing the market to narrate it for you. The ticker becomes optional. The buildings become visible. And the numbers start behaving like explanations instead of surprises.

1.4 Riding the Waves: Market Cycles and Valuations

Real estate has a personality flaw: it responds to yesterday's demand with tomorrow's supply.

When apartments are full and rents are climbing, the industry doesn't say, "Wonderful, let's keep things stable." It says, "Build more." Developers draw plans, lenders write checks, cranes appear, and two or three years later a brand-new wave of supply hits the shoreline. Sometimes it lands perfectly, like a surfer catching the swell at the right second. Other times it lands like a fridge falling off a truck—right when the economy is slowing and tenants are suddenly picky.

That lag is why real estate moves in cycles. Not because buildings are fickle, but because *construction is slow* and *capital has moods*. Demand can change fast. Supply can't. The gap between the two is where booms and busts are born.

The cycle starts with something boring: occupancy

Every property type has its own version of "How full are we?" Apartments call it occupancy. Offices call it leased percentage. Hotels talk about REVPAR and occupancy like sports commentators. Industrial is all about absorption and available space. Different dialects, same idea: when space is scarce, owners have pricing power. When space is abundant, tenants do.

That pricing power shows up first in behavior, not spreadsheets. Leasing teams stop offering freebies. Concessions shrink. "One month free" becomes "maybe we can waive the application fee." Renewal increases get bolder. Tenants grumble but stay because moving is a hassle and alternatives are limited.

Then the market gets a little drunk on its own success.

Overbuilding: the most predictable surprise in real estate

If you ever want proof that humans learn slowly, watch what happens when a property sector becomes profitable. Capital rushes in with the subtlety of a marching band.

The logic is reasonable at the individual level. A developer sees strong rents and thinks, "If I deliver a building in 24 months, I'll be leasing into a great market." A dozen developers think the same thing. The city doesn't get one new building; it gets a cluster of them. The first few lease up quickly. The later ones start competing. Concessions return. Rent growth slows. Someone's underwriting begins to sweat.

The strange part is how *rational* overbuilding feels while it's happening. Every project has a spreadsheet. Every project has a story. By the time the market realizes supply has gotten ahead of demand, the concrete is already cured.

For REIT investors, overbuilding is not a moral failure of "greedy developers." It's a mechanical feature of the business. You don't avoid it by being cynical. You manage it by understanding which property types are easier to oversupply (often those with fewer constraints) and which markets have natural speed bumps (zoning, geography, high construction costs).

Recessions: vacancies don't rise politely

When the economy contracts, real estate doesn't usually break in half overnight. It softens, then weakens, then sometimes snaps in the places that were already strained.

A recession is best understood as a broad decline in economic activity rather than a cute rule like "two negative quarters of GDP." In the real world, recessions show up as hiring freezes, layoffs, delayed expansions, and nervous CFOs. Tenants don't announce, "We are now participating in a recession." They quietly cut square footage, postpone moving, renegotiate leases, or stop traveling. Consumers pull back. Businesses become conservative. And empty space, once rare, starts appearing like weeds.

Vacancy is the problem that turns small issues into large ones because vacancy hits you from two directions:

- you lose revenue (no tenant, no rent), and
- you often spend more to attract replacement tenants (leasing commissions, tenant improvements, marketing, downtime).

This is why real estate cycles can feel unfair. The good times are gradual. The bad times can be quick.

Why real estate cycles create opportunity (and also heartbreak)

Cycles create opportunity because prices don't just reflect today's NOI; they reflect fear, financing costs, and expectations about what NOI will be later. In down markets, buyers demand higher yields to compensate for uncertainty. Sellers either accept lower prices or refuse to sell. Transactions slow. Headlines get dramatic.

That's where disciplined investors tend to make their best future returns: buying when other people are emotionally allergic to the sector. Not because pain is inherently profitable, but because *prices can overshoot to the downside* just as they overshoot to the upside.

The heartbreak comes from confusing "opportunity" with "immediacy." Real estate can stay out of favor for years. A good purchase can look stupid for a while. Markets don't reward you for being early; they reward you for being right *and* being able to wait.

And waiting requires something more valuable than conviction: liquidity and staying power. The investors who capitalize on cycles are rarely the ones with the strongest opinions. They're the ones who can keep holding and keep buying without needing perfect timing.

Valuation in real estate: the Cap Rate

Stocks have price-to-earnings ratios. Bonds have yields. Real estate has the cap rate—short for *capitalization rate*. It's the industry's everyday thermometer for value.

The definition is straightforward:

$$\text{Cap Rate} = \frac{\text{NOI}}{\text{Property Price}}$$

If a property produces $1,000,000 in NOI and sells for $20,000,000, the cap rate is:

$$\frac{1{,}000{,}000}{20{,}000{,}000} = 0.05 = 5\%$$

A cap rate is often described as the property's "yield." That's close enough to be useful, with one important caveat: it's a yield on the property's *current* NOI, not a guaranteed return. Properties have rent growth (or rent decline). They have vacancies. They have expenses that can surprise you. Capex lurks. Cap rates are a snapshot, not a destiny.

Still, cap rates are powerful because they turn the messy physical world into a number that can be compared across deals.

The seesaw relationship: price up, cap rate down

Cap rates have an inverse relationship with price because NOI is the fulcrum. Hold NOI constant and the math forces the seesaw:

- If price rises, NOI/Price falls, so the cap rate goes down.
- If price falls, NOI/Price rises, so the cap rate goes up.

That means a "low cap rate" deal is usually an expensive deal (relative to NOI), and a "high cap rate" deal is usually a cheaper deal (relative to NOI). People sometimes translate this too simplistically into "low cap rate is bad, high cap rate is good." Reality is more interesting.

Low cap rates often belong to properties perceived as safer: great locations, strong tenant demand, stable cash flows, high barriers to new supply. High cap rates often belong to properties perceived as riskier: uncertain demand, lease-up risk, weaker locations, outdated buildings, or markets with plenty of new supply on the way.

In other words: cap rates are not just about return; they're about *risk pricing*. A higher cap rate can be a bargain or a warning label. A lower cap rate can be overpaying or it can be paying up for quality.

Cap rates as an "earnings yield" for buildings

If you're coming from stock investing, cap rates click when you treat them like an earnings yield. A stock's earnings yield is roughly E/P (the inverse of P/E). A property's cap rate is $NOI/Price$. It's the same structure: income divided by price.

That mental model helps with intuition. When investors are willing to accept a lower earnings yield for stocks (higher P/E), we call it "multiple expansion." When investors accept a lower cap rate for real estate, that's also multiple expansion, just with different jargon and better carpeting.

And just like stock multiples, cap rates can move even when property income hasn't changed. That's the part that surprises people. They assume value moves only when NOI moves. Cap rates quietly remind you: valuation is also driven by what investors demand as compensation for risk.

What moves cap rates? Interest rates, expectations, and fear

Cap rates tend to be influenced by the general level of interest rates because investors compare real estate yields to other available yields in the economy. If "safe" rates rise (think Treasury yields), a 4.5% cap rate starts looking less appetizing unless the buyer expects strong growth or unusually low risk. Buyers may respond by demanding a higher cap rate, which mathematically implies a lower price for the same NOI.

But cap rates don't move like a lever connected directly to a single interest rate. They're also shaped by:

- *expected NOI growth* (rent growth, occupancy improvement, expense control),
- *expected value changes* (do buyers think the property will be worth more later?),
- *financing availability* (easy credit can compress cap rates; tight credit can expand them),
- *sentiment* (which is just a polite word for crowd psychology).

Sentiment matters more than people like to admit. In hot markets, buyers pay prices that assume everything will go right. In cold markets, buyers demand pricing that assumes several things will go wrong, just in case. The property hasn't changed. The buyer's stomach has.

A cap rate example that makes the point painfully clear

Take a property with $1,000,000 in NOI.

At a 5% cap rate, its value is:

$$\text{Price} = \frac{NOI}{Cap} = \frac{1{,}000{,}000}{0.05} = 20{,}000{,}000$$

At a 6% cap rate, its value becomes:

$$\frac{1{,}000{,}000}{0.06} \approx 16{,}666{,}667$$

Nothing changed about the building. Same tenants. Same rent checks. Same roof. Same parking lot with the same pothole that everyone pretends not to see. The only change is what buyers require as a yield. That one-percentage-point move reduced the implied value by about $3.33 million, or roughly 17%.

This is why real estate can reprice quickly in public markets. A REIT's share price can drop not because the buildings suddenly got worse, but because the market is applying a higher required yield to the same (or similar) cash flows. It's math wearing the costume of mood.

The two engines of return: income and change in value

Cap rates also help you separate two different return sources:

- *Income return:* the NOI you collect relative to what you paid (the cap rate is a starting proxy).
- *Value change:* what happens if NOI grows (or shrinks) and what happens if cap rates compress (or expand).

In a strong real estate environment, you can get a double tailwind: NOI rises and cap rates fall, pushing prices up faster. In a tougher environment, you can get hit twice: NOI falls and cap rates rise, pushing prices down faster.

Understanding that two-engine setup is calming because it explains why returns can vary so much year to year without requiring a conspiracy. Sometimes the engine of income is fine but the engine of valuation is sputtering. Sometimes it's the opposite. Sometimes both are roaring.

Cap rates and REITs: the public-market funhouse mirror

REIT investors don't buy buildings directly; they buy shares. Shares can swing more violently than private property values because shares trade all day, every day, with zero friction. A pension fund can't sell half an office tower before lunch. A trader can sell $50 million of a REIT in a minute.

That liquidity is a gift and a source of drama. It means REIT prices can temporarily disconnect from private-market cap rates and appraisal values. When fear spikes, public markets often reprice first and ask questions later. When optimism returns, they can reprice faster than private markets can close deals and update comps.

For an investor with patience, that gap can be opportunity. For an investor with a short fuse, it's torture.

The practical takeaway is not "public markets are irrational." It's that REITs carry two kinds of sensitivity:

- sensitivity to the *property business* (NOI, occupancy, rent growth), and
- sensitivity to the *pricing mechanism* (discount rates, required yields, sentiment).

If you only watch one, you'll constantly feel blindsided by the other.

Why cap rates are useful---and why they won't make you a market timer

Cap rates are useful because they force you to anchor on income. They can prevent you from paying luxury prices for mediocre cash flows. They can help you compare sectors (multifamily vs. industrial vs. retail) and markets (a high-demand coastal city vs. a fast-growing Sunbelt market). They can help you sanity-check whether a deal is priced as "safe and growthy" or "risky and stagnant."

But cap rates are not a crystal ball. Real estate can stay expensive longer than you think. It can stay cheap longer than you think. Cap rates can compress in ways that feel irrational when capital is abundant and fear is scarce. They can expand in ways that feel cruel when financing dries up and everyone is suddenly a pessimist.

Valuation measures are best treated like weather forecasts for expected returns: helpful for packing the right clothes, terrible for predicting the exact raindrop that hits your head.

A grounded way to think about cycles as an investor

A cycle-aware investor doesn't need to predict the exact peak or trough. They need to avoid the classic errors:

- extrapolating good times forever,
- assuming bad times are permanent,
- confusing a cap rate move (valuation) with a business collapse (fundamentals),
- and overleveraging so that normal volatility becomes a personal emergency.

Real estate cycles don't care about your timeline. They don't coordinate with your career or your kids' tuition schedule. They just happen. The advantage goes to the person who can keep buying sensibly when prices are interesting and keep holding sensibly when prices are insulting.

Cap rates give you a language for "interesting." Cycles give you a reason prices become interesting at all. Put them together and the market stops feeling like a random number generator and starts feeling like what it is: a system that oscillates between confidence and caution, with rent checks quietly flowing in the background.

Chapter 2

Demystifying the REIT Structure

Imagine owning a slice of a Manhattan skyscraper, a sprawling logistics warehouse, or a local hospital without ever having to fix a leaky toilet or chase down a late rent check. That is the promise of a REIT. But before we go shopping, we need to look under the hood. This chapter is your owner's manual-stripping away the Wall Street jargon to reveal exactly what you are buying, the vast menu of properties available, and the mechanics of how to actually place that trade.

2.1 What Are You Actually Buying?

A REIT share looks like a stock quote and trades like a stock quote, so the brain wants to file it under the same folder as every other public company: "I buy a piece, management runs the business, profits happen, dividends maybe." Close—but not quite. A REIT is less like a typical corporation that *may* share profits and more like a legal-and-tax machine that is *built* to share them. The whole point of the structure is to take a very physical, very lumpy thing (real estate) and turn it into something that can be owned in clean, bite-sized units without forcing the owner to be a plumber, negotiator, or part-time amateur attorney.

The simplest mental model is this: buying a REIT is buying a claim on a *stream*. Not a building. Not a deed. Not a set of keys. A stream. Rent checks (or interest checks), minus expenses, minus financing costs, flowing through a company that has agreed to play by a strict rulebook. That rulebook is what makes the REIT feel different in your account: dividends tend to be more prominent, tax character can be more complicated, and the business can behave like a hybrid between a utility (steady, contractual cash flows) and a financial company (sensitive to rates and capital markets). You are not buying "a property." You are buying *the way money moves through property.*

Trust is the most misunderstood word in "Real Estate Investment Trust." It doesn't mean a family trust where assets sit quietly in a vault. In practice, most REITs are operating companies with employees, strategy, risks, and competition. The "trust" part is a clue that the law intends them to function as a conduit: gather real-estate cash flows, then hand most of the taxable earnings out to the people who provided the capital. In exchange for that generosity, the government demands discipline. A REIT can't wake up one morning and decide to become a software company. It has to stay mostly in its real-estate lane, and it has to pay out most of what it earns in taxable terms.

That last sentence is the hinge. To keep REIT status, the company must distribute at least *90% of its taxable income* to shareholders each year. That one requirement changes the personality of the entire enterprise. A normal C-corporation can treat profits like a pantry: stockpile them, spend them, buy back shares, make acquisitions, or pay a dividend if it feels like it. A REIT's pantry has a wide-open door. Earnings don't linger long. They are pushed outward, usually in quarterly dividends, because the tax advantage depends on being a high-throughput system rather than a savings account.

Here's why that matters: the classic corporation is exposed to *double taxation*. The company earns money, pays corporate income tax, and then shareholders may pay tax again when dividends are distributed. The REIT, by contrast, is designed to be more like a pass-through—not perfectly, not universally, but directionally. When it pays out dividends and meets the rules, it generally gets a deduction for those dividends, which can dramatically reduce (and often largely eliminate) corporate-level income tax on the real-estate income. Think of it as a tax-efficient pipeline: the government allows the flow to move through the company with far less leakage, as long as the company doesn't hoard the water.

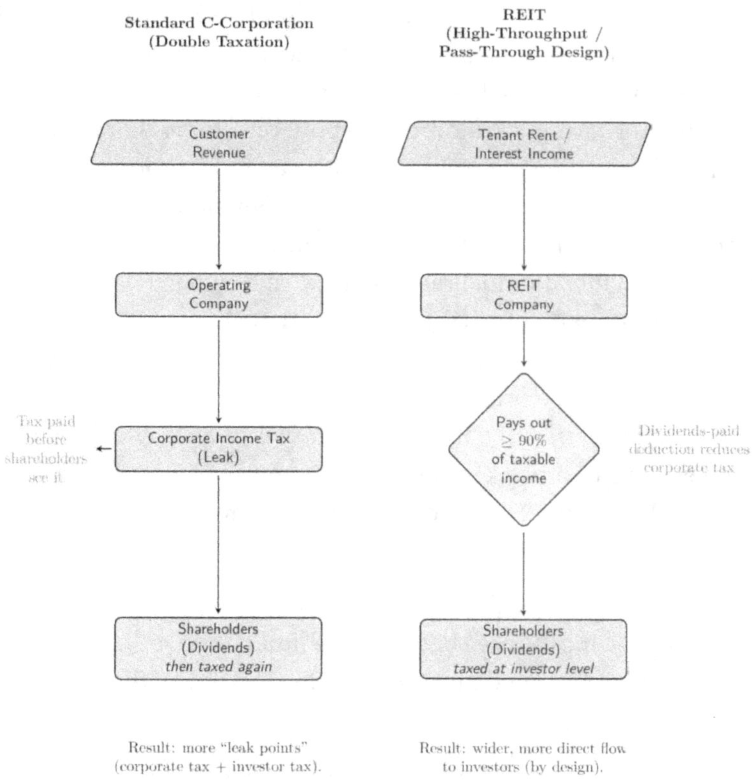

The diagram makes the point visually, but the practical takeaway is even more useful: REITs are often *dividend-forward* because they have to be. That does *not* mean every REIT dividend is safe, or that the dividend is a gift from a benevolent landlord. It's closer to a pressure-release valve. When cash comes in and taxable income is generated, a lot of it must go out. If management wants to grow, it can't rely on quietly accumulating retained earnings the way many corporations do. It has to fund growth with some combination of borrowing, issuing new shares, joint ventures, and asset recycling (selling properties to buy other properties). That funding loop is why REIT investors pay attention to balance sheets and capital markets in a way that, say, toothpaste investors rarely have to.

A subtle but important nuance: the 90% rule is based on *taxable income*, not "cash in the bank." Real estate accounting is full of non-cash expenses, especially depreciation. Depreciation is the tax system's way of saying, "Buildings wear out over time." On paper, that wear-and-tear reduces taxable income even if the property is throwing off steady rent checks. So a REIT can sometimes pay a dividend that is larger than its taxable income without breaking the rules, because the dividend is not limited to taxable income; the rule is about paying *at least* 90% of taxable income. This is one reason REIT tax forms can feel like a drawer full of mismatched socks: part of what you receive may be ordinary dividend income, part may be capital gain distributions, and part may be *return of capital* (which typically reduces your cost basis rather than being taxed immediately). The cash can be simple. The tax character can be anything but.

That complexity is not a reason to avoid REITs; it's a reason to respect what you own. A REIT is a tax and cash-flow machine built on top of real assets. It produces both income and paperwork. If you want the convenience of owning real estate in stock form, the paperwork is the cover charge.

Now zoom in on the business itself. When you buy a broad-market index fund, you're buying a slice of the economy. When you buy a REIT, you're buying a slice of a much narrower universe with very specific physics. A property is a long-lived asset with contracts attached (leases) and financing attached (mortgages, unsecured bonds, credit facilities). Those two attachments are where the drama lives.

Leases mean the revenue line isn't a daily popularity contest. Tenants sign agreements. Many leases include annual rent bumps. Some leases push expenses onto the tenant. That can make cash flows feel pleasantly predictable compared to businesses that must resell their product every day. But leases also come with their own set of hazards: a lease is only as good as the tenant, and a building is only as valuable as its next tenant. Real estate is where "location, location, location" is true—and also incomplete. The hidden sequel is "financing, financing, financing."

Financing matters because real estate is expensive, and REITs generally use leverage. Leverage isn't automatically reckless; it's often the tool that makes returns possible on stabilized assets. But it means a REIT lives with interest rates in the background like a bass note. Rising rates can increase borrowing costs, compress property values, and make investors demand higher yields (which often pressures share prices). Falling rates can do the opposite. Either way, REITs are rarely indifferent to the cost of capital.

So what exactly is a REIT share in your hands? Economically, it's a claim on:

- *Net operating cash flow* from properties or mortgages,
- *The residual value* of those assets after debts are paid,
- *Management's skill* at allocating capital in a world where growth must be financed externally more often than not.

That last bullet is why two REITs in the same sector can feel like different species. One management team plays defense, keeps leverage modest, and treats the dividend like fine china. Another pushes growth, runs closer to the edge, and treats the dividend like a marketing budget. The structure is the same; the temperament is not.

At this point the word "REIT" can still feel like a single thing. It isn't. There are two major families that share the same acronym but behave very differently in storms: *Equity REITs* and *Mortgage REITs*.

Confusing them is like confusing a building owner with a bank because both are involved in real estate.

An *Equity REIT* is the landlord. It owns properties—apartment buildings, warehouses, data centers, self-storage facilities, shopping centers, cell towers, timberland—and its core job is to turn square footage into durable cash flow. If you picture a landlord collecting rent, paying for maintenance, paying property taxes, and improving the building to keep tenants happy, you're in the right neighborhood. The risks tend to be operational and economic: occupancy, rent growth, tenant health, property-level expenses, local supply and demand, and the ever-present question of whether the building will still be desirable ten years from now.

A *Mortgage REIT* (often shortened to mREIT) is the banker. It owns *real-estate-related debt* rather than the buildings themselves. That debt might be residential mortgage-backed securities, commercial mortgages, or other structured credit instruments. The cash flow comes from *interest income* and the spread between what the mREIT earns on its assets and what it pays to finance them. If that sounds more like a financial firm than a property company, that's because it is. The main risks lean heavily toward the world of finance: interest rate movements, funding conditions, credit spreads, hedging effectiveness, and the stability of short-term borrowing markets.

Both types can pay big dividends. For a beginner, that fact is a trap. Big dividends can come from the calm productivity of occupied buildings, or they can come from a levered portfolio of bonds with razor-thin margins and enormous sensitivity to rates. The dividend yield alone won't tell you which one you're looking at. It's like judging a car's safety by how loud the engine is.

A good first-pass question is embarrassingly simple: *What is the REIT actually holding?* If the answer is "properties," you're dealing primarily with leases, tenants, and real estate operations. If the answer is "mortgages" or "mortgage-backed securities," you're dealing primarily with interest rates, leverage, and market plumbing. There is no shame in preferring one over the other, but it's wise to admit what you're signing up for. Most people who want "real estate exposure" in a portfolio are picturing the property-owner version. They want the landlord economics without the landlord headaches. That's generally the equity REIT lane.

That distinction also explains why the language around REIT performance can be confusing. Equity REIT investors talk about occupancy, same-store net operating income, lease spreads, development pipelines, and property types. Mortgage REIT investors talk about book value, net interest margin, duration, convexity, prepayment speeds, and hedges. Those are not just different vocabularies. They are different realities.

Even within equity REITs, you still aren't buying a building in the way a direct owner buys a building. Direct ownership is painfully specific. It's one property (or a small handful), in one place, with one set of tenants, one roof, one set of local regulations, and one disaster waiting patiently on the calendar. A REIT is a portfolio business: dozens, hundreds, sometimes thousands of properties. Your risk is diluted across a map rather than concentrated in a single address. That diversification is not magic—entire sectors can struggle at once—but it changes the shape of risk. A surprise vacancy is an annoyance to a large REIT. It can be a personal financial event to a small landlord.

But there's a trade. Direct owners can choose when to sell, when to refinance, when to renovate, and which tenant to accept. A REIT shareholder can't call management and request a kitchen remodel. Your control is traded away for liquidity and scale. You can exit with a click, but you can't micromanage. That sounds obvious until you realize how many people buy REITs while secretly craving the feeling of control that comes with physical property. A REIT will not scratch that itch. What it offers instead is something most landlords would happily buy if it existed in a bottle: *effortless partial ownership, priced continuously, diversified automatically.*

The tax angle is where the "what am I buying?" question becomes more than philosophy. A REIT dividend is often taxed differently than the qualified dividends many investors are used to from typical U.S. corporations. In plain English: a REIT dividend is frequently more like *rental income passing through* than like a slice of corporate profit that gets preferential rates. Many REIT dividends are taxed at ordinary income rates (though portions can be capital gains or return of capital, and tax law can provide partial deductions for certain REIT dividends). That can make REITs feel "less tax-efficient" in a regular taxable account compared to broad-market equity funds, even when the underlying business is excellent.

This is not a flaw in REITs; it's the price tag on the bargain. The REIT structure is designed to reduce taxation at the corporate level, not necessarily to make every shareholder's tax bill small and tidy. The government is essentially saying: "We'll tax this income once, mostly at the investor level, as it flows out." Compare that to the normal corporation: "We'll tax it at the company, then potentially again at the investor." Different leaks, different places.

That also explains why REITs are often discussed alongside tax-advantaged accounts. Putting REITs in an IRA or other sheltered account can keep the ordinary-income feel of many REIT dividends from constantly tugging at your taxable return each year. Whether that's best for you depends on your overall portfolio, tax situation, and time horizon, but the core concept is sturdy: the REIT is built to distribute, and the tax system treats much of what it distributes as income-like.

One more important layer: a REIT can look like it's paying you "income" while simultaneously reinvesting heavily. How? Because real estate is capital-hungry, and the REIT's growth engine is often external financing. A high payout does not necessarily mean a REIT is starving itself. It may be recycling capital: issuing shares to buy properties that increase total cash flow, using debt prudently, and selling mature assets to fund higher-growth ones. This is why REIT analysis leans on cash-flow measures that attempt to capture sustainable earning power after the real-world costs of maintaining buildings.

If that sounds like a lot of moving parts for something marketed as a simple dividend stock, that's the point: REITs *are* simple in one sense—you can buy them like any other stock—but the underlying machine has gears. Those gears are not there to confuse you. They're there because turning rent into dividends at scale is a craft, and the law forces the cash to keep moving.

So, what are you actually buying?

You're buying a share of a legally constrained real-estate business whose tax advantages come with an agreement: stay mostly in real estate, and send most taxable earnings out the door. You're buying exposure to leases, local supply and demand, financing conditions, and management's capital allocation skill. If it's an equity REIT, you're buying the landlord's economics without the landlord's chores. If it's a mortgage REIT, you're buying the banker's economics without the banker's office—and with most of the banker's risks.

The best part is that you don't need a deed to benefit from real estate's superpower: the ability to turn a physical, useful asset into a long-running cash stream. The important part is knowing which stream you purchased. Rent stream or interest stream. Wide pipe with modest leaks, or a system where the leaks can move around. A ticker symbol can hide those differences. Your job as an investor is to refuse to let it.

2.2 A Tour of Property Sectors

Real estate has a branding problem. Say the words out loud and most people picture either an apartment building with a leaky lobby fountain or a glass office tower with a lonely coffee shop on the ground floor. That mental image is understandable—and wildly incomplete. Public REITs own pieces of modern life that don't look like "real estate" until you remember the unromantic definition: *land plus long-lived, income-producing stuff bolted to it*. By that definition, a warehouse that keeps your online order moving, a server farm that stores your photos, and a forest managed for timber are all playing the same game: turn a physical footprint into predictable cash flow.

Sector labels are the vocabulary investors use to avoid accidentally buying the wrong weather. Not the wrong stock—the wrong *weather*. Some property types thrive when consumers splurge; others don't care. Some love inflation because leases reprice quickly; others are stuck in long contracts that move slowly. Some behave like steady toll roads; others swing like cyclical businesses wearing a landlord costume. The most useful way to think about sectors is to ask: *What force actually pays the rent here?* A paycheck? A hospital admission? A data plan? A corporate IT budget? A moving truck and a life transition? That driver is the engine under the hood.

REIT sector tours often turn into laundry lists. The problem with a laundry list is that everything looks equally important, and nothing sticks. A better approach is to group sectors the way the economy experiences them: the digital systems you barely notice, the industrial plumbing you only notice when it breaks, the demographic realities nobody votes on, and the traditional street-corner properties that make cities feel like cities. Same legal wrapper, different heartbeat.

The Digital Backbone: Data Centers and Cell Towers

If you want a fast way to recalibrate your idea of "property," start with a data center. From the outside it can look like a bland warehouse with unusually serious fencing. Inside is a high-density ecosystem of servers, cooling systems, backup generators, and network connections that turn electricity into rent checks. Tenants aren't browsing for a cute location near a park. They're buying uptime, connectivity, and redundancy—the modern equivalents of water pressure and reliable power.

Data center REITs sit at the intersection of real estate and infrastructure. The buildings matter, but so does the ecosystem around the building: fiber routes, power availability, and the ability to add capacity without melting the equipment. Leases can be relatively sticky because moving critical IT workloads is expensive and risky. But the sector carries a unique kind of obsolescence risk: the building has to keep up with changing technology and tenant needs. A beautiful 1998 office building can still be an office building; a poorly designed data center can become a very expensive concrete box.

Cell tower REITs are even more unintuitive. A tower is, in real-estate terms, a very skinny building. The rent comes from wireless carriers leasing space on the structure for antennas and related equipment. One tenant becomes two, two becomes three, and suddenly the economics look like a high-margin apartment building where adding a new "resident" barely increases operating costs. These leases often have built-in escalators, and the underlying demand driver is simple: more devices, more data, more network density.

The charm of digital-infrastructure REITs is that the demand feels structural rather than trendy. People can argue about the next app; they don't argue about the direction of data usage. The risk is that the tenant base can be concentrated (a handful of carriers, a handful of hyperscale cloud companies), and the assets can be politically and technologically entangled. Zoning, permitting, power grids, and local opposition can matter as much as national GDP.

The E-Commerce Engine: Industrial and Logistics

Industrial REITs own the buildings that make modern consumption possible: warehouses, distribution centers, logistics hubs, and "last-mile" facilities closer to dense populations. These properties are the opposite of glamorous, which is part of their appeal. They don't need to be charming. They need to be efficient boxes in the right places.

The economic driver here isn't retail sales in general; it's *the movement of goods*. When more shopping shifts online, inventory needs a home, and that home needs to be close enough to deliver quickly. Even traditional retailers increasingly behave like logistics companies with branding budgets. A good warehouse near a major metro can be more valuable than a fancy store on a secondary street because it sits closer to the real bottleneck: time.

Industrial leases often have annual rent escalators and can be structured as "triple-net" (meaning the tenant pays many property-level expenses like taxes, insurance, and maintenance). That tends to make cash flow steadier, but it also means tenant credit matters. You don't want a beautiful building with a tenant whose balance sheet looks like a Jenga tower. Another quirk: new supply can arrive quickly. It's easier to build a warehouse than a downtown skyscraper, so hot industrial markets can attract developers like moths to a porch light. The sector can feel steady right up until it doesn't, especially if demand slows and new supply lands at the same time.

Still, industrial is a terrific example of how REITs let you own a theme without picking a retail winner. You don't have to guess which brand dominates next year. You own the shelves, the loading docks, and the geography.

Demographic Plays: Health Care and Senior Housing

Some sectors are tied less to the economy and more to biology. Health care real estate is not a bet on medical innovation; it's a bet on the fact that people will keep aging, needing care, and showing up for appointments whether or not consumer confidence is feeling optimistic.

Health care REITs can own a mix of property types: medical office buildings (where outpatient care happens), hospitals (more complex, often with specialized operators), skilled nursing facilities, and senior housing. The sector is a masterclass in the difference between a property and an *operating business*. A medical office building leased to doctors looks like real estate with relatively stable tenants and long-ish leases. Senior housing can look more like hospitality: staffing, occupancy, pricing, and resident turnover create operating sensitivity that feels closer to running a service business than collecting rent from a warehouse.

That operating component is why health care REITs can be misunderstood. Investors sometimes expect all health care real estate to behave like a bond: steady and boring. Then a labor shortage hits, expenses spike, and margins compress. The building didn't change. The economics did. In this corner of REIT-land, the quality of the operator can matter as much as the quality of the property.

The long-term tailwind is demographic. The short-term reality is that health care is heavily regulated, reimbursement systems change, and public policy can drift like a weather front. Owning the real estate does not immunize you from the system that pays the tenants.

Residential: Apartments, Single-Family Rentals, and Manufactured Housing

Residential REITs own housing, which sounds straightforward until you notice how many different housing businesses hide under the same umbrella. Apartments in coastal cities. Sunbelt garden-style communities. Single-family rental portfolios. Manufactured housing communities. Student housing. Each has its own rhythm.

Apartments are often praised for their ability to "reset" rents relatively quickly. In many markets, leases are about a year long. That means rent growth can reflect inflation and local wage growth faster than in property types with decade-long leases. The flip side is that apartments feel economic stress quickly too. If renters pull back, concessions show up fast, renewals soften, and occupancy becomes a weekly conversation.

Single-family rental REITs are a different animal. They own scattered-site houses, often in suburban markets. The demand driver is household formation and the cost (or impossibility) of buying. Operating costs can be higher on a per-unit basis because maintenance is spread out, but tenant retention can be strong when families treat the home as a long-term base. This segment also brings a social dimension: housing is emotional, and large landlords can become political lightning rods.

Manufactured housing communities are one of the most counterintuitive sectors for new investors. The business often involves owning the land and infrastructure (roads, utility hookups) while residents own their homes and pay lot rent. That can create sticky tenancy because moving a manufactured home is expensive and disruptive. The sector can produce durable cash flow and steady demand, but it also sits in the middle of debates about affordability and tenant protections. It's not just a spreadsheet; it's a lived-in asset class.

Residential REITs, more than most, teach the core REIT lesson: the property type is the headline, but the lease structure and tenant dynamics write the plot.

Retail: Not Just Malls (and Not All Malls Are the Same)

Retail is where outdated stereotypes go to die. "Retail REIT" does not automatically mean "dead mall." The sector spans grocery-anchored shopping centers, open-air lifestyle centers, freestanding net-lease properties, and yes, enclosed malls. These are different businesses with different tenant mixes, lease structures, and sensitivities.

A grocery-anchored center tends to behave like a convenience engine. People buy food in good times and bad. That stable traffic can support smaller tenants like pharmacies, quick-service restaurants, and service providers. The risks are local competition, tenant quality, and consumer spending, but the format has proven resilient because it aligns with errands rather than entertainment.

Enclosed malls are more polarized. The best malls can be cash machines with high-end tenants, tourism, and limited local competition. The weak malls can spiral as anchors leave and foot traffic drops. The sector is less about "malls" and more about *real estate quality and tenant productivity*. A mall is a platform; some platforms print money, others become expensive nostalgia.

Net-lease retail REITs deserve special mention because they are often marketed as sleepy income vehicles. They own freestanding properties leased to tenants on long leases where the tenant pays many expenses. Think of a drugstore, convenience store, or big-box retailer. The appeal is predictable rent and minimal property management. The risk is concentration and lease duration. Long leases can be comforting—until inflation rises or interest rates move and you realize your rent bumps are small and your contract is stubborn.

Retail teaches humility. Consumers change. Formats change. Locations that feel unbeatable can fade. The best retail owners are part landlord, part anthropologist.

Office: The Sector Everyone Has an Opinion About

Office REITs are a reminder that a lease is not the same thing as a guarantee. Office demand is tied to how companies organize work, and that is a cultural choice as much as an economic one. A strong office building in a prime location with modern amenities can still attract tenants. A commodity building with outdated layouts can struggle even if the rent looks cheap.

What makes office particularly tricky is that leases can be long, which delays the feedback loop. Weak demand doesn't always show up immediately in cash flow. It shows up as higher leasing costs, tenant improvement spending, and tougher renewal negotiations. The building can look stable right before a wave of expirations forces reality to the surface.

For investors, office can be a case study in why sector labels are only the first filter. "Office" is not one market. It's many micro-markets: CBD trophy towers, suburban campuses, life-science-adjacent clusters, secondary cities, government-heavy regions. You're not buying "office." You're buying *this* office, in *this* employment ecosystem, with *this* tenant roster and *this* balance sheet.

Self-Storage: Real Estate Powered by Life Transitions

Self-storage sounds like a punchline until you look at the business drivers. People rent storage units when they move, divorce, downsize, inherit furniture, start a small business, renovate, or simply run out of space. In other words, storage demand is tied to life events rather than purely to economic cycles.

Storage properties can be relatively cheap to build compared to many other property types, and they can be run with lean staffing. Pricing can adjust quickly because leases are short and customers are often month-to-month. That flexibility can be a superpower in inflationary periods. It can also make the sector more competitive: if returns look attractive, new supply can appear, and pricing power can soften.

The fascinating part of storage is behavioral. Customers value convenience and inertia. Many intend to rent for a month and stay for a year. A well-located facility can become a quiet annuity built on human procrastination. It's not romantic, but it's very real.

Hotels and Lodging: The REIT Sector with the Shortest Memory

Hotels are the opposite of long-lease real estate. A hotel "lease" is a night. That means revenue resets constantly. When travel demand is strong, hotels can raise rates immediately and cash flow can surge. When demand weakens, the drop can be swift and dramatic. Lodging REITs can behave less like landlords and more like cyclical operating businesses that happen to own valuable real estate.

This sector is often driven by a mix of business travel, leisure travel, convention calendars, airline capacity, and consumer confidence. It can be exhilarating. It can also be punishing if you expected the steady cadence of apartment rent checks. For diversification, lodging can add a different kind of return stream, but it's not a beginner's idea of "safe income."

Specialty and Alternative Sectors: Timberland and More

Some REIT sectors feel like they were invented to test whether investors are paying attention. Timberland REITs own forests and manage them as renewable resources. The cash flow comes from harvesting timber and selling wood products, and the asset itself grows biologically over time. That creates a unique flexibility: if prices are unattractive, trees can sometimes be left to grow, effectively "storing" value on the stump. It's real estate with a living component.

Other specialty categories exist in the broader REIT ecosystem (and classifications can evolve), but the point stands: the REIT wrapper can hold far more than the classic trio of office, mall, and apartment. The constraint is not imagination; it's whether the underlying assets and income fit the rules and the economics make sense in public markets.

A Practical Way to Use Sectors Without Turning It Into Homework Hell

Sector knowledge is only useful if it changes decisions. The goal isn't to memorize a taxonomy. The goal is to avoid accidental concentration and to align your REIT holdings with the kind of risk you can live with.

One practical habit is to treat sector exposure like ingredients rather than finished meals. If your portfolio already leans heavily on the consumer (retail stocks, consumer discretionary companies, travel), piling into lodging REITs might not diversify as much as it seems. If your career income is tied to tech budgets, owning only data centers and towers can quietly stack the same macro risk in a different outfit. If you crave stability and inflation-linked rent growth, apartments, self-storage, and some forms of necessity retail can behave differently than long-lease net-lease properties. None of these are rules. They're guardrails.

Another habit: don't confuse a sector's headline story with a specific REIT's balance sheet and lease book. Two companies can operate in the same property type and have radically different risk profiles based on tenant concentration, lease duration, development exposure, and leverage. Sector is the genre. The balance sheet is the plot twist.

The most satisfying part of REIT investing is that you can own a small slice of the physical economy with intentionality. Want the hidden infrastructure behind the internet? There are REITs for that. Want the warehouses that make two-day shipping possible? There are REITs for that. Want housing exposure without a midnight call about a broken water heater? There are REITs for that too. "Real estate" stops being a monolith and starts looking like a menu—and once you see the menu, it becomes much harder to accidentally order the same dish over and over.

2.3 Ways to Access the Market

Buying REITs is seductively easy. Your brokerage app doesn't make you pass a quiz on lease structures or debt maturities. It just offers a button that says *Buy*. That convenience is a gift and a trap. It's a gift because it turns real estate into something liquid and scalable. It's a trap because the method you choose to get exposure—a single REIT stock, a broad ETF, or a private non-traded product with glossy brochures—quietly determines what kinds of mistakes you're even *able* to make.

The market offers three main doors. One door gives you control. One door gives you convenience. One door often gives you a sales pitch. They can all lead to real estate exposure, but the fee drag, liquidity, and transparency on the other side feel like three different worlds.

Door #1: Buying Individual REIT Stocks (Control With a Homework Bill)

Choosing individual REITs is the closest you can get to being a real-estate "owner" while still staying in the clean world of ticker symbols. You pick the property type, you pick the balance sheet style, you pick the management team you want to bet on. If you're the kind of person who reads restaurant reviews before ordering take-out, stock-picking REITs can be oddly satisfying: you're not buying "real estate," you're buying *this* landlord, with *this* tenant roster, in *these* markets, funded in *this* way.

The biggest advantage is precision. Want logistics warehouses but not retail? Want apartments in one region but not another? Want a conservative balance sheet and minimal development risk? You can curate those preferences rather than accepting whatever a fund happens to hold. Over time, that ability to avoid sectors you don't understand (or simply don't want) can matter as much as the ability to pick winners.

The second advantage is cost. Buying a stock doesn't come with an explicit management fee the way a fund does. Your broker might charge nothing for the trade, and the REIT's internal operating costs are already reflected in its financials. That doesn't mean stock-picking is "free"—it means the costs are less visible. Your main cost becomes behavioral: the temptation to trade too often, to chase yield, or to abandon a good business because the share price had a bad month.

The obvious downside is concentration risk, and it's not just about owning too few names. It's about owning too few *economic engines*. You can build a portfolio of five REITs that looks diversified because the tickers are different, while all five are quietly exposed to the same force (say, higher interest rates, or a regional job market, or a single tenant category). A REIT is a funnel for cash flows, and funnels can be correlated in ways that aren't apparent from the logos.

There's also the competence problem. Some investors hear "public real estate" and assume it's easier to analyze than a typical company because buildings are tangible. In reality, a REIT can be simple or it can be a Swiss watch. Lease structures, tenant credit, refinancing calendars, development pipelines, and property-level capex (capital expenditures, meaning money spent to maintain or improve assets) can make two REITs in the same sector behave very differently. If you don't want to spend time learning those gears, stock picking will eventually invoice you in the currency of unpleasant surprises.

Door #2: REIT ETFs and Funds (Convenience With a Small, Persistent Toll)

A REIT ETF is the fast-food combo meal of real estate exposure: instant diversification, predictable packaging, and no need to meet the chef. It's a genuinely good deal for many investors, especially early on, because it solves the hardest part of REIT investing with one purchase: avoiding accidental overconfidence.

Diversification is not just about owning many REITs. It's about owning many *property types and lease models* at once. When one sector hits an air pocket, another may be cruising. A broad REIT ETF can spread your exposure across landlords whose tenants pay rent for very different reasons: housing needs, data usage, medical services, shipping and logistics, and everyday shopping. That variety can smooth the ride without requiring you to become an expert in every sector.

ETFs also reduce the "single-name blowup" problem. If one REIT stumbles—bad acquisition, tenant bankruptcy, refinancing trouble—a diversified fund turns that into a paper cut instead of a broken bone. That's not a guarantee of safety (sector-wide declines still happen), but it changes the odds of a catastrophic self-inflicted wound.

The cost is the expense ratio: a small annual fee charged by the fund to cover management and operations. Expense ratios can be modest, but they are relentless. They don't care if your portfolio is up or down; they take their slice regardless. Think of it as a tiny hole poked in your return bucket. One hole may not matter much. Over decades, it's worth respecting.

The other cost is imprecision. A broad REIT ETF buys what the index tells it to buy. That can mean exposure to sectors you dislike, or to companies you would never choose on purpose. Some funds are sector-specific (industrial-only, data-center-heavy, etc.), which can be useful, but then you're back to making sector bets. A fund doesn't eliminate decisions; it changes which decisions you're making.

ETFs can also lull investors into ignoring what they own. A single REIT invites scrutiny because it has a story. A fund feels anonymous, like a utility bill. That anonymity is dangerous if it stops you from noticing concentrations (for example, a heavy weight in one or two mega-REITs or in one fast-growing sector). The fund wrapper doesn't erase risk; it distributes it.

Door #3: Private and Non-Traded REITs (The Sales Pitch Door)

Here's where the tone changes. Publicly traded REITs and ETFs are sold in the open market. Prices move every day, and liquidity is immediate. Private or non-traded REITs often arrive through a different channel: a broker, an advisor, a seminar, a webinar, or a beautifully designed packet that sounds like it was written by someone who moonlights as a poet.

These products often emphasize three phrases that feel like emotional relief:

- "Not correlated to the stock market"
- "Stable income"
- "Less volatility"

All three can contain a grain of truth. All three can also be deeply misleading, depending on what the product is actually doing.

Start with the volatility claim. Non-traded REITs can *appear* less volatile because you don't see a price flicker every second. That's not the same as being less risky. It's the difference between a bumpy road and a car with a dashboard that only updates the speed once a month. The bumps still exist. You're just not watching them in real time.

Liquidity is the real dividing line, and it's where many investors get blindsided. A publicly traded REIT can be sold on any trading day (subject to market conditions). A non-traded REIT may have limited redemption programs with restrictions, waiting periods, gates, and the ability to suspend redemptions. If you need your money back quickly, "you own real estate" turns into "you own a promise about when you might be able to sell."

Fees are the second dividing line, and they can be brutal in private real estate products. Public REIT costs are mostly visible in financial statements and market pricing. In non-traded structures, fees can be layered: upfront selling commissions, ongoing management fees, acquisition fees, disposition fees, and sometimes incentive structures that reward growth in assets under management more than growth in per-share value. You don't need to assume bad intentions to understand the risk: a high-fee structure must run faster just to stand still.

Transparency is the third dividing line. Public REITs file detailed financials and are relentlessly judged by the market. Non-traded products can be more opaque, with valuations derived from appraisals and internal estimates rather than continuous market discovery. That doesn't automatically make them fraudulent or worthless. It does mean you are relying more heavily on the sponsor's reporting and governance. In investing, "trust me" is not an asset class.

A particularly confusing subset is the "public non-traded" REIT: registered with regulators but not listed on an exchange. These are not the same as tiny private deals, and they can be legitimate vehicles.

They can also still carry the core frictions: limited liquidity, complex fees, and pricing that doesn't behave like a live market. Some newer structures use net asset value (NAV) reporting and periodic repurchase programs, which can improve clarity compared to older non-traded models. Even then, the central trade remains: you're exchanging day-to-day market liquidity for a controlled exit door.

None of this means every non-traded REIT is automatically wrong for every investor. It means the burden of proof shifts. A public REIT starts with the presumption of liquidity and transparent pricing. A non-traded REIT starts with questions:

- How do I get out, exactly, and under what conditions can that door close?
- How much am I paying in total, including fees that don't show up as a tidy expense ratio?
- How is the valuation determined, and how often is it updated?
- Who benefits most when new money comes in: existing shareholders, or the sponsor?

If a product can't answer those questions with plain language and written specifics, it's not an investment; it's a fog machine.

A Decision Matrix You Can Actually Use

The cleanest way to compare the doors is to judge them on three axes that tend to determine investor outcomes more than any brochure ever will: *liquidity, fees, and transparency*. Liquidity tells you whether you control your timeline. Fees tell you how much headwind you're accepting. Transparency tells you whether you can verify what you own without faith.

Place the matrix near any conversation about private REITs, because private products are where investors most often confuse "less visible" with "less risky."

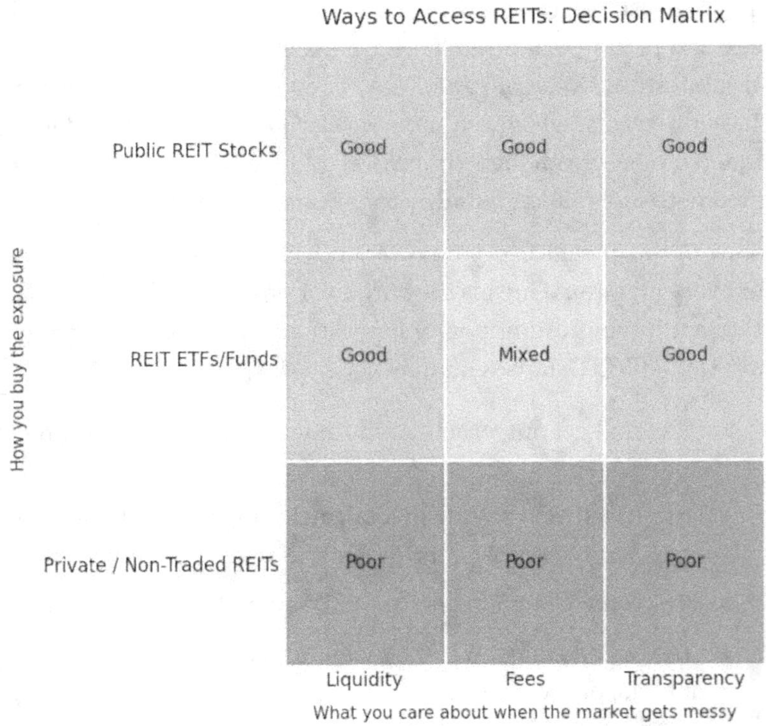

Green = generally investor-friendly. Yellow = trade-off. Red = common pain point.

The matrix isn't a moral judgment; it's a friction map. Public REIT stocks and ETFs score well because markets force them to be liquid and well-lit. Private and non-traded REITs score poorly on liquidity and fees in the general case because friction is part of their design. They may be trying to stabilize capital or fund long-duration projects without daily price pressure, but that stability comes at the cost of investor flexibility and, often, investor bargaining power.

So Which Door Should You Use? A Reality-Based Answer

If you enjoy analyzing businesses and you're willing to read the boring parts, individual REIT stocks can be a great fit. You get control, you avoid fund fees, and you can shape your exposure intentionally.

The price of admission is doing enough work to know what you own and why the dividend exists.

If you want broad exposure without turning REITs into a second job, a REIT ETF is the most straightforward tool on the shelf. You're paying a small fee to outsource diversification and reduce single-company risk. That's often a fair trade, especially when your main edge is consistency rather than cleverness.

If you're being pitched a non-traded REIT as a volatility cure, treat it like someone offering to sell you a "stress-free" treadmill. Maybe it's well-built. Maybe it's useful. But you should be intensely curious about what's being hidden: the exit restrictions, the fee stack, and the valuation method. A product that can't tolerate scrutiny is not built for your benefit.

The underlying irony is that many investors are drawn to private REITs because they want the calm of real estate without the mood swings of the stock market. Public REITs already offer that calm at the *business* level—leases, rent checks, physical assets. The stock price wiggles because markets wiggle. Paying extra fees and surrendering liquidity just to avoid seeing the wiggles is an expensive form of emotional management.

Liquidity is not just a convenience. It's a form of safety. It's the ability to admit you were wrong, to rebalance when life changes, and to raise cash without begging for permission. In REIT investing, the best access method is usually the one that keeps you in control of your own timeline while keeping the fee drag from quietly eating your returns in the dark.

2.4 Accounts and Mechanics

A REIT can be a beautifully engineered cash-flow machine, and you can still lose money by clicking the wrong button. Not because the market is out to get you, but because the stock market is a marketplace, and marketplaces have friction: spreads, settlement rules, tax buckets, and tiny fees that only feel tiny until you add them up for ten years. Owning REITs "the right way" isn't about having secret information. It's about avoiding avoidable damage.

The good news: the mechanics are learnable in an afternoon. The even better news: once you internalize them, you stop feeling like the market is a casino and start feeling like it's a hardware store. You walk in with a list, you pay a fair price, you carry your purchase home without dropping it, and you don't accidentally buy the deluxe version of something you didn't need.

Step One: Choose the Account That Matches the Job

A brokerage account is just a container with rules. The investments are the furniture; the account is the house. Many people spend all their time debating which couch to buy and forget to check whether the house has a roof.

Most investors will own REITs in one (or both) of these containers:

- a standard taxable brokerage account, where dividends and capital gains can create taxes along the way,
- a tax-advantaged retirement account (like an IRA), where taxes are deferred or structured differently.

The difference matters more for REITs than for many other stock investments because REIT distributions often arrive with a heavier "ordinary income" flavor. When a distribution is taxed as ordinary income, it gets taxed like wages, not like a preferential long-term qualified dividend. That isn't a reason to fear REITs; it's a reason to match the asset to the account.

A practical way to decide is to ask: *Do I want this holding primarily for income right now, or for long-term compounding?* In a taxable account, high-yield holdings can feel productive while quietly generating a steady tax bill. In a tax-advantaged account, the same income can be reinvested without the yearly tax drag. The trade-off is flexibility:

taxable accounts are easier to access without retirement rules, while retirement accounts are built for the long game and come with constraints.

If you're building a long-term REIT allocation and you have room in a retirement account, it's often a natural home for REITs. If you need the liquidity of a taxable account, REITs can still belong there—just go in with eyes open and a willingness to keep records clean.

Step Two: Pick a Brokerage Like You're Picking Plumbing

A brokerage is not a personality test. It's plumbing. You want clean execution, reliable recordkeeping, and low unnecessary fees. The design of the app and the cleverness of the marketing are not the product; the product is whether you can place a trade without surprises and whether tax forms arrive correctly in January.

A few features matter more than people expect:

- *Fractional shares* (helpful if you want to invest a fixed dollar amount regularly).
- *Dividend reinvestment* (often called DRIP), so distributions can automatically buy more shares.
- *Good cost-basis tracking* (critical for REITs because return-of-capital distributions can affect your basis).
- *Clear order-entry tools* (limit orders, time-in-force options, and visibility into bid/ask quotes).

If your brokerage makes it hard to place a limit order, or hides the bid-ask spread behind a cute animation, it's not doing you favors. You're not there to be entertained.

Step Three: Understand What Happens After You Click *Buy* ($T + 1$ Settlement)

When you place a trade, two things happen: *execution* and *settlement*. Execution is the moment your order matches with someone else's order. Settlement is when the money and the shares officially exchange hands behind the scenes.

In U.S. markets, most stocks (including REITs) now settle on $T + 1$, meaning one business day after the trade date. If you buy on Monday, settlement is typically Tuesday. This matters because your broker may show the position immediately, but the back-office exchange of cash and securities is on that timeline.

Why should you care? Because settlement rules are where beginners accidentally break things:

- Selling a security and immediately buying another with the proceeds can be fine in many accounts, but it can create issues in cash accounts if you trade faster than settlement.
- If you're moving money in and out, settlement timing affects when cash is truly available.

It's not complicated once you know it exists. The mistake is assuming the market works like a debit card. It doesn't. It works like a check that clears quickly, but not instantly.

Step Four: Market Orders vs. Limit Orders (The ``Don't Tip the Market" Rule)

A market order is a command that says, "Buy (or sell) immediately at the best available price." It feels decisive. It also hands your pricing power to the market at that exact moment. Most of the time in very liquid stocks, that's not a big deal. In thinner-traded REITs, it can be an expensive habit.

The hidden cost is the *bid-ask spread*. The *bid* is the highest price someone is currently willing to pay. The *ask* is the lowest price someone is currently willing to sell for. The gap between them is the market's toll for immediacy.

If a REIT has a bid of $49.90 and an ask of $50.10, the spread is $0.20. That doesn't sound like much. But if you place a market buy, you often pay near the ask. If you later sell with a market order, you often receive near the bid. Congratulations: you just paid the spread twice. Do that repeatedly and you've created your own personal expense ratio, except it's invisible and it doesn't come with diversification.

A limit order is your antidote. A limit order says, "Buy, but only at $X or better," or "Sell, but only at $Y or better." It turns you from a price taker into a price negotiator. You may not get filled instantly, and that's the point: you are refusing to overpay for the privilege of speed.

A simple operating rule for REITs:

- Use *limit orders* by default.
- Use *market orders* only when liquidity is deep and the spread is tight (and even then, check the quote).

If you want a number to anchor your judgment, look at the spread as a percentage of the price. A $0.02 spread on a $100 stock is a rounding error. A $0.30 spread on a $10 REIT is a meaningful haircut.

Time-in-force matters too. A *day* limit order expires at the end of the trading day if it doesn't fill. A *good-til-canceled* (GTC) order can sit open longer, which is convenient, but it also means you can forget it exists and accidentally buy on a sudden dip you didn't intend to catch. Convenience is always a double-edged sword.

Step Five: Dividends Have Dates (and the Calendar Can Bite You)

REIT investing attracts dividend-focused investors, and dividend-focused investors are often tempted to do something that sounds clever and usually isn't: buy right before the dividend date to "capture" the dividend.

Dividends come with a small set of dates:

- the *declaration date* (the board announces the dividend),
- the *ex-dividend date* (the first day the stock trades without the upcoming dividend attached),
- the *record date* (the company checks who owns shares),
- the *pay date* (cash is paid).

The market adjusts for dividends. When a stock goes ex-dividend, the price often drops by roughly the dividend amount, all else equal. You didn't create money; you shifted it from "price" to "cash." The only consistent winners in dividend-capture games are taxes and spreads.

The smarter use of dividend mechanics is operational, not tactical:

- Make sure you're enrolled (or not enrolled) in dividend reinvestment intentionally.
- Keep cash available if you plan to reinvest manually with limit orders.
- Avoid panic if the price drops around the ex-date; that's usually arithmetic, not betrayal.

Step Six: Know What Your 1099-DIV Is Trying to Tell You

REIT distributions are famous for showing up on tax forms looking like a mixed bag. That's not a clerical error; it's the tax system acknowledging that REIT payouts can represent different underlying economic sources.

In a taxable account, your broker will generally report REIT distributions on Form 1099-DIV, and those distributions may be split into categories such as:

- *ordinary dividends* (often the main component for REITs),
- *capital gain distributions* (taxed more like long-term capital gains),
- *nondividend distributions / return of capital* (often reduces your cost basis rather than being taxed immediately).

This matters because two REITs with the same cash dividend yield can produce different tax outcomes. It also matters because return of capital is not "free money." It's often a deferral mechanism that lowers your basis, which can increase your taxable gain when you eventually sell.

Your job is not to memorize tax code. Your job is to keep your records accurate and to understand the basic translation: *the cash you received is not always taxed the same way.* If you invest through a retirement account, much of this complexity is muted because the account's tax rules dominate. In a taxable account, it's part of the ongoing maintenance.

There's also the topic of the Section 199A-related reporting that can appear for REIT dividends (often shown in a dedicated box on the 1099-DIV). Tax law changes can affect how this works over time, and it's an area where outdated internet advice ages badly. The practical investor move is simple: keep the tax form, use the figures reported, and avoid making decisions based on a benefit you can't verify on your actual paperwork.

Step Seven: Reinvestment Choices (Automatic DRIP vs. Manual Rebuy)

Dividend reinvestment is where mechanics become behavior. A DRIP automatically uses your dividend to buy more shares, often on the pay date. It's frictionless compounding, and it can be a great default if you want to build a position steadily without thinking about it.

The drawback is price control. Automatic reinvestment typically buys at the market price (or a broker-determined price) when the reinvestment happens. If the REIT is thinly traded or the spread is wide, you can wind up buying at slightly unfavorable prices repeatedly. It's not catastrophic, but it's real.

Manual reinvestment gives you control: you can pool dividends as cash and place limit orders at prices you consider reasonable. The drawback is that manual reinvestment requires you to actually do it. Many people don't. Cash piles up, plans evaporate, and the compounding you imagined becomes the compounding you postponed.

A blunt rule that respects human nature:

- If you are not reliably hands-on, DRIP is often better than good intentions.
- If you are hands-on and the REIT is illiquid, manual reinvestment with limit orders can reduce spread costs.

Step Eight: A Simple Execution Checklist

Before you place a REIT trade, run a quick mental checklist. It sounds small, but small mechanics are where most retail investors leak returns:

- *Account*: Is this in taxable or retirement, and am I okay with how dividends will be treated there?
- *Liquidity check*: What's today's volume and the current bid-ask spread?
- *Order type*: Limit order, not market, unless the spread is meaningfully tight.
- *Position size*: If this goes down 20%, will I regret the size more than the price?
- *Reinvestment plan*: DRIP on, or cash for manual buys?

None of this predicts the market. It just reduces the number of ways you can accidentally pay too much, sell too cheaply, or tax yourself unnecessarily.

REIT investing is often marketed as a way to own real estate without landlord headaches. The fine print is that you trade landlord headaches for investor mechanics. The mechanics are gentler, cleaner, and far less likely to involve midnight phone calls. But they still matter. The investor who respects them gets what REITs are best at delivering: boring consistency at the operational level, and compounding that doesn't get quietly mugged by avoidable friction.

Chapter 3

Analyzing Quality and Value

Imagine trying to judge a runner's speed by measuring how much water they drink. That is exactly what happens when you use standard stock metrics, like the P/E ratio, to judge Real Estate Investment Trusts. In the world of real estate, accounting rules can make a profitable landlord look like a money-loser and a risky bet look like a safe harbor. In this chapter, we swap the wrong ruler for the right one. We will learn the language of landlords-translating accounting noise into true cash flow-so you can confidently spot the difference between a bargain and a value trap.

3.1 Decoding REIT Financials

If you've ever looked at a REIT's income statement and thought, *How is this thing paying a dividend when it barely earns money?*, congratulations: you've just met the most common optical illusion in real estate accounting.

GAAP net income—the bottom-line "profit" number most investors instinctively trust—was built for businesses where factories rust, trucks wear out, and software becomes obsolete on schedule. Real estate refuses to play along. A well-located apartment building can be *more* valuable at age 30 than at age 3, yet GAAP treats it like a melting ice cube: every year, a slice of "depreciation" gets deducted as if the building is steadily disintegrating into dust.

That's not just an academic quirk. Depreciation can be the single largest expense on a REIT's income statement, and it's largely non-cash—meaning it reduces reported earnings without necessarily reducing the dollars landing in the REIT's bank account. The result is a recurring phenomenon: perfectly healthy REITs that look mediocre (or even unprofitable) on paper, especially if you judge them with the usual stock-market lenses like earnings per share and P/E ratios.

Here's the twist that makes depreciation especially mischievous in real estate: GAAP depreciates *the building* but ignores *the land*. Land often appreciates. Buildings can be renovated. Rents can rise. Neighborhoods can improve. Meanwhile, the accounting keeps marching the "value" of the structure downward as if time alone is a wrecking ball.

Depreciation isn't *fake*—a roof really does age, elevators really do break, and lobbies really do get shabby. The problem is that GAAP depreciation is a blunt instrument: it applies a schedule, not a reality check. It doesn't ask whether the property is in higher demand today than when it was purchased. It doesn't care that a landlord can push rent on renewal while the depreciation expense stays the same. It's a rule, not a lens.

The practical consequence is simple: if you rely on net income to judge a REIT, you can end up penalizing the best kind of real estate—the kind that quietly becomes irreplaceable.

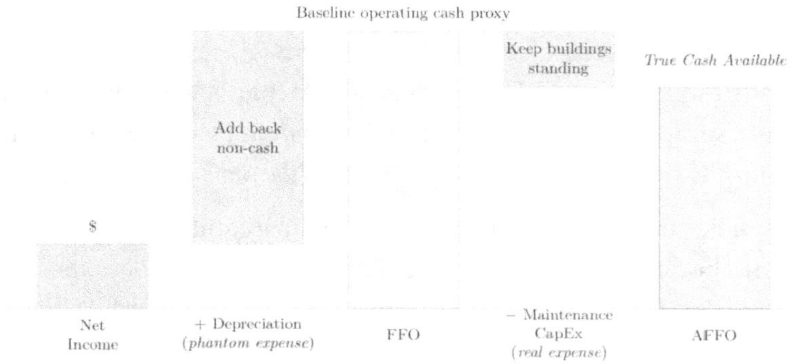

Now for the vocabulary that REIT investors use when they want the story to match the economics.

Funds From Operations (*FFO*) is the industry's most widely used adjustment to GAAP net income. Think of it as a "real estate translation layer." The basic idea is straightforward: start with net income, then add back depreciation and amortization related to real estate, because those expenses often say more about accounting rules than about cash leaving the building.

FFO also typically strips out certain gains or losses from property sales. Why? Because selling a building can create a one-time accounting pop (or drop) that has nothing to do with the ongoing earning power of the remaining portfolio. A REIT can look brilliant for a year because it sold one asset at a big gain—then look ordinary the next year when that gain disappears. FFO tries to keep your attention on the recurring machine, not the occasional fireworks.

A good mental model is to imagine a REIT as a basket of properties that throws off rent checks. The rents pay the property operating bills (taxes, insurance, repairs, payroll), and what's left helps cover corporate overhead, interest, and dividends. Net income is that story after GAAP adds its own dramatic soundtrack—especially depreciation. FFO lowers the volume on the soundtrack.

But FFO still isn't the number that pays the dividend. Cash does. And buildings, unlike fairy tales, require upkeep.

That's where *Adjusted Funds From Operations* (*AFFO*) enters. AFFO is often described as a REIT's "cash available for distribution," and the most useful way to think about it is: FFO, *minus the recurring spending required to keep the portfolio earning rents at the same level.*

The most common deduction is maintenance capital expenditures—often shortened to maintenance CapEx. This is money spent on things that prevent the property from sliding backward: replacing worn carpet, resurfacing a parking lot, updating HVAC components, repairing a roof section, refreshing unit interiors, and so on. These are real checks written to real contractors. Ignore them long enough and the property doesn't just look tired—it starts losing tenants, and then the income statement gets ugly in a way no adjustment can hide.

Here's the key nuance that keeps investors honest: depreciation is a *phantom* expense (usually non-cash in the current period), while maintenance CapEx is a *real* expense (cash leaving the business). GAAP mixes these together into a single "profit" narrative. FFO removes the phantom. AFFO tries to subtract the real.

That difference is the reason two REITs can both report the same FFO, yet one has a safer dividend than the other. One portfolio might be newer, requiring less ongoing maintenance. Another might have older assets that demand constant reinvestment just to stay competitive. FFO can treat them like twins; AFFO starts to show which one is quietly paying a "keep-it-running tax."

A simple numerical example makes this feel less like jargon and more like common sense. Imagine a REIT reports:

- GAAP net income that looks modest, because depreciation is heavy.
- Strong occupancy and steady rent collections.

- A dividend that seems too big relative to earnings.

That's the classic moment where FFO and AFFO prevent a bad conclusion. The REIT might be throwing off plenty of cash; it's just being "charged" an accounting expense that doesn't reflect a current cash outflow. The more important question becomes: after setting aside the money required to keep the properties in shape, how much cash remains to pay shareholders?

A practical way to keep the concepts straight is to treat each metric like a different camera lens:

Metric	*Best used for*
GAAP Net Income	Legal/accounting baseline; spotting unusual items; understanding how conservative (or volatile) reported earnings are.
FFO	Normalizing operating performance for real estate's depreciation distortion; comparing REITs on a more apples-to-apples basis.
AFFO	Approximating recurring, distributable cash after routine property upkeep; pressure-testing dividend capacity.

Notice what's missing: none of these is "the one true number." They're tools. Real estate is too lumpy and too physical for a single metric to tell the whole story.

Also, a word of caution that's worth taking personally: FFO is widely used, but it's not GAAP. AFFO is even less standardized. That doesn't make them shady—it just means you have to read like an adult, not like a headline scanner.

Companies can be disciplined and conservative in how they calculate AFFO, subtracting a realistic estimate of recurring capital needs. Others can be... optimistic. The most common "optimism leak" is pretending that maintenance CapEx is tiny, or calling clearly recurring expenses "non-recurring adjustments." When you see a REIT tout "record AFFO" while the properties are aging and the tenant base is demanding upgrades, skepticism is not cynicism—it's basic maintenance.

So how do you spot whether an AFFO number is built for clarity or built for marketing?

Start with the reconciliation. A well-behaved REIT doesn't just toss "AFFO" onto a slide like it's a law of physics. It shows the bridge from net income to FFO and then to AFFO, with line items you can recognize and argue with. When the adjustments multiply like rabbits—especially with vague labels like "other normalized items"—you've earned the right to slow down.

Next, sanity-check maintenance CapEx. There's no universal "correct" number because property types differ. A self-storage facility and a hotel are both real estate, but one is basically a row of metal boxes and the other is a living, breathing machine with furniture, fixtures, and constant wear. The point isn't to memorize a single percentage. The point is to ask: *Does this deduction look believable for this kind of property?*

Then, look for consistency over time. A REIT that reports near-zero recurring CapEx year after year is implicitly claiming its buildings are perpetually young. That can be true for short windows—especially after a wave of new development—but real estate eventually sends invoices.

Another nuance: AFFO is often discussed as "take-home pay," but even that phrase has fine print. Some REITs also adjust for straight-line rent (an accounting smoothing technique), leasing costs, or recurring tenant improvements. Those can be reasonable adjustments—leases often require upfront cash to secure long-term rent streams—but they also create room for interpretation. Interpretation is where comparability goes to die.

The cure is not to avoid AFFO. The cure is to treat it like a conversation, not a commandment. If you can explain each adjustment in one plain-English sentence, you're probably on solid ground. If you can't, you're not "missing something"—you're being asked to trust something.

One more important distinction keeps investors from mixing up property-level health with company-level health: *Net Operating Income* (*NOI*) versus FFO/AFFO.

NOI is a property-level metric—rent and other property income minus property operating expenses. It ignores corporate overhead, interest expense, and how the REIT is financed. That makes it great for valuing buildings (because a building doesn't come with a corporate HQ attached), but incomplete for valuing a REIT (because shareholders *do* inherit the financing choices).

FFO and AFFO live at the REIT level. They reflect the whole machine: properties, overhead, and the capital structure that can either amplify returns or turn a mild downturn into a dividend headache.

This is why a REIT can brag about strong same-property NOI growth while still having a stressed dividend. The properties might be fine. The balance sheet might be the issue. Metrics don't just measure performance; they reveal *where* the risk is hiding.

And yes—GAAP still matters.

It's tempting to treat net income as "useless" once you learn FFO, but that's an overcorrection. GAAP can surface real economic pain that FFO and AFFO might smooth over: impairment charges (which can be a sign management overpaid), persistent interest expense pressure, or a pattern of gains from asset sales that masks weak organic performance. In other words, GAAP can be the smoke alarm. FFO is the floor plan. AFFO is the budget after you've paid the bills that keep the lights on.

A final way to make all of this feel intuitive is to picture two landlords.

Landlord A owns a well-located apartment building that's getting more desirable each year. The neighborhood improves, rents rise, and the building is cared for. GAAP still "depreciates" the structure annually, pushing reported earnings down—even though the landlord's cash flow may be stable or rising.

Landlord B owns an aging property that's slowly losing its competitive edge. The depreciation expense on paper might look similar, but the *cash* reality is different: more repairs, more turnover, more concessions to keep tenants. If you only look at net income, both landlords can look similarly "unprofitable." If you look at AFFO-like cash after maintenance, the gap shows up where it belongs: in the ability to keep paying out cash without starving the buildings.

That's the heart of REIT financial analysis. The goal isn't to find a magical metric that replaces thinking. The goal is to stop being misled by an accounting rule that treats prime real estate like a disposable appliance—and to start asking the question that actually matters:

After the phantom expenses are stripped away, and after the real-world upkeep is paid for, how much cash is left for you?

3.2 Ensuring Dividend Safety

A high REIT yield can mean "income machine." It can also mean "distress flare." The market isn't generous for no reason. When you see an 8% or 10% dividend yield, assume you're being handed a message in a bottle. Your job is to read it before you cash the checks.

Start with the uncomfortable truth: dividend cuts are rarely lightning bolts from a clear sky. They're usually slow-motion math problems—cash comes in, obligations go out, and somewhere in the middle management realizes the payout has turned into a weekly game of financial Jenga. By the time the press release lands, the balance sheet has often been whispering the ending for quarters.

The basic mindset shift is simple but powerful: don't ask whether the dividend looks *attractive*. Ask whether it looks *defensible*. Attractive is a number. Defensible is a system.

Step one: confirm what the dividend is actually competing against

A REIT dividend isn't the only claim on cash. It's competing with interest expense, debt maturities, property upkeep, and the constant temptation (or necessity) to refinance at whatever rate the bond market feels like offering that morning. A dividend can be "covered" in a calm year and still be fragile in a stressful one if the REIT is boxed in by near-term obligations.

The fastest way to avoid being seduced by yield is to reframe it: a dividend is a promise to keep sending cash out the door *even when* the world gets inconvenient. If the REIT's cash is already spoken for by lenders and maintenance needs, the dividend becomes the easiest lever to pull—because unlike the mortgage payment, it's optional.

So the checklist begins with coverage, but it cannot end there.

Step two: use the payout ratio that matches the business

For REITs, the payout ratio that matters is not the traditional "dividends as a percent of earnings." Earnings can be a hall of mirrors. What you want is a cash-flow proxy that reflects ongoing reality—how much cash the portfolio throws off after routine costs.

That's why many REIT investors focus on the *AFFO payout ratio*:

$$\textit{AFFO payout ratio} = \frac{\text{Dividends paid}}{\text{AFFO}}$$

Read it like a household budget. If your household brings home $5,000 a month and you commit $4,000 to a fixed payment, you're not "rich." You're *tightly scheduled*. One surprise medical bill and the credit card comes out. A REIT paying out nearly all of its AFFO is living the same way.

A payout ratio in a reasonable band can be boring—in the best sense. It suggests the REIT has room to handle a rough patch, fund recurring investment, and still mail the dividend. A payout ratio consistently above 100% is the opposite of boring. It means the REIT is paying more than it's sustainably generating, which forces it to bridge the gap with asset sales, new debt, or issuing new shares. That's not "income." That's financing.

There's a subtle but crucial point here: *a REIT can legally pay a dividend even if it doesn't have the cash.* It can borrow. It can sell properties. It can issue shares. The dividend check can still clear—right up until it can't. The payout ratio is how you catch the tension early.

A trap disguised as a spreadsheet: ``adjusted" numbers that adjust away reality

AFFO is incredibly useful, and it's also incredibly easy to abuse.

A well-run REIT explains its adjustments clearly and consistently, quarter after quarter, like a person who keeps honest books. A more promotional REIT can turn "AFFO" into a choose-your-own-adventure story, where recurring costs are labeled "one-time," and inconvenient expenses are treated like they happened on another planet.

Here's the investor's litmus test: if the payout ratio looks miraculously safe only because the REIT keeps adding back expenses that sound recurring in plain English—leasing costs, tenant improvements, normal repair cycles—then you may be looking at a dividend supported by vocabulary, not cash.

You don't need to be an accountant to spot this. You just need to be stubborn about definitions. If an item keeps showing up, it's not non-recurring. It's *the business*.

Step three: watch the dividend's ``cushion," not just its coverage

A payout ratio isn't a pass/fail exam; it's a measure of breathing room. Two REITs can both cover the dividend today, yet one is living with a cushion and the other is living with a countdown.

The cushion is the portion of recurring cash flow that remains after the dividend is paid. That leftover cash does more than sit there looking pretty. It funds small upgrades that prevent tenant turnover. It helps refinance costs without begging the equity market. It provides optionality—one of the rarest and most valuable assets in real estate.

If the REIT pays out nearly everything, management's options shrink fast. When a tenant leaves, when a lender tightens terms, when a storm forces capital spending earlier than planned, the dividend becomes the only flexible expense left. The cut isn't always caused by collapsing rents; sometimes it's caused by *no slack*.

Step four: read the debt maturity schedule like a weather forecast

Dividend investors often treat debt like background scenery: it's there, it's big, but it doesn't feel immediate. That's a mistake. Debt is where "safe yield" quietly turns into "yield trap," especially in periods when refinancing gets more expensive.

The debt maturity schedule answers one blunt question: *When does this REIT have to go back to the market and ask for money?*

If maturities are staggered over many years, the REIT is less likely to be forced into refinancing during a bad window. If maturities cluster in the next one to three years, the REIT can be fine operationally and still face a cash crunch—because it has to pay off or refinance debt whether rates are friendly or not.

Think of maturities as a set of doors that must be opened on specific dates. Some doors open into sunny refinancing markets. Some open into a hallway full of lenders with clipboards asking for higher spreads, tighter covenants, and less patience. A dividend funded by yesterday's interest rate environment may not survive tomorrow's.

This is also why the *weighted average debt maturity* is a useful summary statistic. Longer average maturities don't guarantee safety, but short maturities are an unforced error in a business where assets are long-lived and refinancing cycles can be brutal.

Step five: look for interest-rate shock absorbers

Debt isn't just about *when* it's due. It's also about *how* it behaves as rates move.

A REIT with mostly fixed-rate, long-term debt can often ride out a rate spike without immediate damage to cash flow. A REIT with a lot of floating-rate debt feels higher rates quickly, like a thermostat turned up in a locked room. Interest expense rises, AFFO gets squeezed, and suddenly a dividend that looked fine last year is fighting for oxygen.

You don't need to predict interest rates to use this. You just need to recognize sensitivity. If the REIT has meaningful variable-rate exposure, ask: how much of the dividend depends on rates staying cooperative?

Many REITs use hedges (like interest rate swaps) to reduce floating-rate exposure. Hedges can help, but they're not magic. They expire. They cost money. They can provide time—but time is only valuable if management uses it to improve the balance sheet rather than extend the illusion.

Step six: stress-test the dividend against a tenant reality check

A dividend is ultimately paid by tenants. Tenants don't care about your yield. They care about their own finances and alternatives.

That's why "dividend safety" always circles back to the portfolio's operating resilience. A few operating questions cut through noise:

- Are occupancy and same-store performance stable, or drifting the wrong way?
- Is rent growth driven by genuine demand, or by temporary factors like short-term leases that happen to reprice quickly?
- Are tenants high quality (creditworthy, sticky, essential), or are they the first to disappear in a slowdown?

Some property types naturally have more cash-flow volatility. Hotels can reprice nightly and swing hard with the economy. Office can suffer slow-motion occupancy erosion that looks manageable until lease rollovers hit. Healthcare can depend on operator health and reimbursement dynamics. None of this makes a REIT "bad," but it changes what "safe payout ratio" should mean. Volatility demands cushion.

When a REIT runs a tight payout ratio in a volatile sector, the dividend may be covered on average but fragile at the edges—exactly where real-life downturns live.

Step seven: separate taxable rules from investor reality

REITs are often described with a catchy phrase: "they must pay out 90%." The phrase is directionally true and dangerously incomplete.

The legal rule is tied to *taxable income,* not GAAP net income, and not your favorite cash-flow proxy. Taxable income has its own set of deductions, timing differences, and quirks. That means a REIT can be obligated to distribute cash even in situations where cash flow feels tight, and it can sometimes distribute more (or less) than a casual observer expects.

The practical investor takeaway is not "ignore the rule." It's this: don't assume that because a REIT pays a dividend, the dividend is therefore safe. A REIT can be a very law-abiding entity and still be a very stressed business.

Step eight: check whether the dividend is funded by the business-or by the capital markets

A sustainable dividend is paid from recurring cash generation. An *unstable* dividend is often paid from external financing—issuing shares, selling assets, or drawing on credit.

One of the cleanest warning signs is a REIT that pays a fat dividend while repeatedly issuing equity at unattractive prices. That's like a restaurant owner handing you a coupon funded by selling slices of the building. The coupon feels great. The economics are ugly.

Asset sales can also mask problems. A REIT can sell properties to fund dividends for a while, especially if it owns good assets. But selling the better properties to fund the dividend is like selling the reliable car to pay the monthly bus pass. It buys time while shrinking future earning power.

So whenever you see a high yield, ask a blunt question: *If the capital markets slammed shut for 18 months, could this dividend still be paid?* If the answer is "no," then the yield is not just a measure of income—it's a measure of dependence.

Step nine: look for a management team that treats the dividend like a reputation asset

A dividend policy is corporate personality made visible.

Some management teams treat the dividend as a reputation asset—something to defend through cycles, even if it means slowing growth, selling non-core assets, or being conservative when times are good. Others treat it as a marketing tool—something to maximize until the math breaks, at which point they "reset" and move on.

You can often tell which type you're dealing with by reading their language. Do they explain dividend decisions in terms of long-term coverage and balance-sheet strength, or do they focus on the headline yield as if it's a product feature? Do they talk about funding needs and reinvestment, or do they talk like dividends fall from the sky?

Then look at the history. A REIT that has maintained or grown dividends thoughtfully across difficult periods has earned a sliver of trust. A REIT that repeatedly swings between generous payouts and abrupt cuts is telling you it values flexibility for itself more than stability for you.

Step ten: use a ``cut story" checklist-because cuts have a pattern

Dividend cuts typically come from a small set of repeatable stories. You don't need to predict which one will happen; you just need to identify whether the ingredients are in the kitchen.

- *Coverage story:* AFFO per share slips, payout ratio rises, cushion disappears.
- *Refinancing story:* large maturities approach, borrowing costs jump, cash flow gets reallocated from dividends to lenders.
- *Portfolio story:* occupancy deteriorates or lease rollovers reset rents downward, shrinking cash flow faster than expenses can adjust.

These stories can overlap, and they often do. The common thread is that the dividend is rarely the first thing to break—it's the last. Management will cut capital spending, slow acquisitions, sell assets, tap the revolver, issue equity, renegotiate covenants. The dividend gets cut when the other levers have become too painful or too visible.

That's why dividend safety analysis is about early warnings. By the time the dividend looks obviously unsafe, the stock price has usually done the math already.

A practical ``sleep well" routine for REIT dividends

If you only remember one habit, make it this: whenever a yield makes your eyes widen, force yourself to do a quick, boring audit. Excitement is the enemy of underwriting.

Question	*What a good answer feels like*
Is the dividend covered by recurring cash flow (AFFO)?	Coverage with a cushion; not a razor-thin payout ratio that leaves no room for surprises.
When does the REIT need to refinance debt?	No scary maturity wall in the near term; a

None of this eliminates risk. Real estate is cyclical, tenants surprise you, and capital markets can change moods faster than a cat.

What this checklist does is change the kind of risk you take. Instead of unknowingly buying a yield that's high because it's endangered, you start buying dividends with a fighting chance—dividends backed by cash flow, protected by time, and supported by balance sheets that don't require the world to stay pleasant.

A safe REIT dividend isn't the one with the highest yield. It's the one that can keep paying when nobody's in the mood to refinance, tenants are negotiating harder, and management has to prove that the distribution is more than a number on a slide deck.

3.3 Valuation Frameworks

A REIT can be a wonderful business and a terrible stock at the wrong price. That sounds obvious until you live it: you buy a "high-quality" landlord, the buildings perform, rents climb, the dividend shows up like clockwork—yet the share price goes nowhere for years because you paid a luxury multiple on day one. Real estate has a special talent for making this mistake feel *reasonable*. The properties are tangible. The story is comforting. The valuation is where the trap door hides.

Two valuation lenses do most of the heavy lifting in REIT investing. One asks, *What are the buildings worth if you had to price them like real estate?* That's Net Asset Value (NAV). The other asks, *What am I paying for the REIT's cash-generating power compared with similar REITs?* That's relative multiples, usually Price-to-FFO (P/FFO) or Price-to-AFFO (P/AFFO). Neither is perfect. Together, they stop you from confusing "familiar" with "cheap."

NAV: pricing the REIT like a pile of buildings (because it is)

NAV is the closest thing REITs have to an "appraisal-based" anchor. The concept is almost suspiciously plain:

$$NAV \approx \text{Market value of real estate} - \text{Net liabilities}$$

If you could sell every property at a fair price, pay off the debts, settle the obligations, and sweep what's left into a bucket for shareholders, that bucket is the rough idea behind NAV.

The reason investors love NAV is that it speaks the native language of real estate: property values, not accounting earnings. It's a reminder that a REIT is not just a dividend stream-it's ownership of assets that can be worth more (or less) than the stock market currently implies.

The reason investors get hurt with NAV is that it feels more precise than it is. NAV is not printed on a stone tablet. It's an estimate built from assumptions, and the biggest assumption has teeth.

Cap rates: the small percentage sign that moves mountains

Most NAV work starts by translating a property's Net Operating Income (NOI) into a market value using a capitalization rate (cap rate). The cap rate is simply:

$$\begin{aligned} \textit{Cap rate} &= \frac{\text{NOI}}{\text{Property value}} \\ \Rightarrow \quad \text{Property value} &= \frac{\text{NOI}}{\textit{Cap rate}} \end{aligned}$$

That rearranged equation is where the drama lives. Divide by a smaller number, and the value jumps. Divide by a bigger number, and the value falls. Cap rates are the market's way of saying, "Given the risk and growth outlook, this is the return I demand from this building."

Now watch what that means in practice. Suppose a property produces \$10 million of NOI.

- At a 5% cap rate, value is \$200 million.

- At a 6% cap rate, value is $166.7 million.

Nothing about the building's bricks changed. The rent checks didn't magically shrink. But a one-point cap rate move drops the value by about 17%. Scale that across a REIT portfolio and you can see why REIT stocks can swing violently when interest rates, credit conditions, or sentiment shift. NAV is not just "what the properties are worth." It's what the market would pay *today* for the income those properties produce.

Cap rates also vary by property type and quality. A warehouse leased long-term to a strong tenant in a logistics hub does not trade like a struggling suburban office building with near-term lease rollovers. Treating cap rates as a single universal number is like pricing every used car as "a car." NAV requires you to respect the differences.

From property value to NAV: the bridge that investors forget to cross

Once you estimate the market value of the real estate, you still have to get to what shareholders actually own. Debt sits in the middle like a toll booth.

A simplified NAV per share framework looks like this:

$$NAV\ per\ share \approx \frac{(\sum \text{Property values}) + \text{Other assets} - \text{Total liabilities}}{\text{Shares outstanding}}$$

There are details-preferred equity, joint ventures, development pipelines, cash, intangible lease assets, and so on. But the soul of the calculation is constant: *assets minus claims, divided by shares.*

And here's the part that should make you sit up: two REITs can own very similar real estate, yet have very different NAV outcomes because one is conservatively financed and the other is loaded with leverage. Debt doesn't just magnify returns; it magnifies the consequences of being wrong about asset values. When cap rates move against you, leverage turns a valuation haircut into a scalp.

Premiums and discounts to NAV: the market's opinion, not the market's mistake

Once you have an NAV estimate, you compare it to the stock price. This is where REIT investors start using bargain language—"discount!"—like they've just found designer clothing on a clearance rack.

The math:

$$\textit{Discount (or premium) to NAV} = \frac{\text{Share price} - \textit{NAV per share}}{\textit{NAV per share}}$$

If a REIT trades below NAV, it's at a discount. If it trades above, it's at a premium.

The tempting conclusion is: discount equals cheap, premium equals expensive. Real life is meaner. Sometimes a discount is a gift. Sometimes it's a diagnosis.

A REIT can trade at a discount for reasons that should make you cautious:

- The market believes the properties are worth less than your NAV estimate (cap rates should be higher, NOI is fragile, occupancy is rolling over).
- The balance sheet is tight, and future refinancing will siphon cash away from growth and dividends.
- Management has a history of value-destroying capital allocation (buying high, selling low, issuing shares when the stock is cheap).

A premium also has two faces. It can be froth—investors overpaying for a fashionable sector. Or it can be earned: a REIT with excellent assets, low-cost capital, and a track record of turning acquisitions and development into real per-share growth may deserve to trade above what a static appraisal suggests.

A helpful way to think about it: NAV is a snapshot of *assets today*. The stock price is the market's attempt to price *assets today plus the future decisions around those assets*. If management has historically turned one dollar of retained cash and external capital into more than a dollar of value, the market may rationally pay a premium. If management tends to light money on fire, the market will demand a discount as protection.

The ``warranted" premium: when management is part of the asset

Investors often treat management like background noise—people in suits who appear on earnings calls and then vanish. In REITs, management can be a compounding machine or a compounding problem.

A high-quality management team with access to low-cost capital can buy or build properties at returns that exceed its funding costs, repeatedly. That spread becomes growth. Over time, the ability to execute accretive deals is valuable enough that it can justify a premium to NAV—because NAV alone doesn't fully capture the value of the platform.

The opposite is also true. A REIT can own decent properties yet trade at a persistent discount because investors don't trust management to protect shareholder value. In that case, the discount is not irrational pessimism; it's an insurance premium the market charges you for governance risk.

NAV tells you what the castle is worth. The premium or discount often tells you what investors think of the people holding the keys.

Relative multiples: pricing the cash engine, not the appraisal

NAV is asset-centric. Multiples are cash-flow-centric.

For REITs, the most common multiple is Price-to-FFO:

$$P/FFO = \frac{\text{Share price}}{\text{FFO per share}}$$

Think of P/FFO as the REIT version of a P/E ratio-except it's built on a real estate-friendly earnings proxy. P/AFFO does the same thing with a measure closer to recurring cash available after maintenance spending:

$$P/AFFO = \frac{\text{Share price}}{\text{AFFO per share}}$$

Multiples are incredibly useful because they are fast, comparable, and naturally connected to returns. If you pay a lower multiple for the same quality of cash flow, your future return tends to improve (all else equal). If you pay a high multiple, you are implicitly assuming growth will bail you out.

That last phrase—*growth will bail you out*—is where investors get themselves into trouble. A high multiple is not a moral failing. It's a forecast. It says: *I believe this REIT will grow per-share cash flow strongly and reliably, and that the cash flow deserves a premium because it's durable.*

If that forecast is wrong, the multiple compresses, and you can lose money even while the REIT is still performing operationally.

What ``cheap" looks like in multiples (and why it might be a mirage)

A low P/FFO can mean you've found an underappreciated gem. It can also mean the market expects the FFO to fall, or believes the assets are deteriorating, or fears refinancing pain, or anticipates equity dilution. Multiples don't just reflect value; they reflect doubt.

One of the most common ways investors get fooled is by assuming today's FFO is a stable baseline. In some sectors, it isn't. Hotels can swing with travel demand. Mortgage REITs (a different animal entirely) can see earnings shift with yield curves and spreads. Even within equity REITs, a portfolio with heavy lease rollover in a weak market can see NOI reprice downward over time. A low multiple may be "cheap" only if the numerator is low *and* the denominator is trustworthy.

A simple gut-check question helps: *Is this multiple low because the stock price fell, or because the cash flow is at risk?* Sometimes it's both. The difference matters.

How to use comparables without turning your brain off

Relative valuation works best when the peers are truly comparable: similar property type, similar lease structure, similar leverage, similar growth profile. Comparing an industrial REIT with long-term leases to a hotel REIT is like comparing a bond to a roulette wheel and calling both "income."

Even within the same property type, capital structure can distort multiples. A more levered REIT may show higher FFO per share in good times, making it look "cheaper" on P/FFO—right up until the leverage bites. Multiples are not immune to financial engineering.

When multiples feel confusing, it often helps to anchor them to a narrative about risk:

- *Higher warranted multiple* often corresponds to steadier cash flows, stronger balance sheet, better growth runway, and cheaper cost of capital.
- *Lower warranted multiple* often corresponds to higher cyclicality, tenant risk, near-term refinancing pressure, uncertain asset values, or weaker governance.

The multiple is a summary of what investors fear and what they're willing to pay to avoid those fears.

Putting NAV and multiples together: two flashlights, different angles

NAV and multiples sometimes disagree. That disagreement is not a problem; it's information.

- A REIT can trade at a big discount to NAV but also at a low P/FFO. That often means the market thinks the NAV is stale or optimistic, or that the cash flows are about to weaken, or that leverage will force value transfer from shareholders to lenders.
- A REIT can trade at a premium to NAV and a high multiple. That can mean the market expects strong growth and trusts management to compound value, or it can mean investors are overconfident and pricing in perfection.

When both NAV and multiples point to "cheap," you may have a genuine bargain—or you may have a well-lit value trap. The difference is usually found in the boring places: lease rollover schedules, tenant health, debt maturity ladders, and the realism of capital spending assumptions. Valuation doesn't replace underwriting; it forces underwriting to be honest.

A compact valuation worksheet you can actually use

When you want a quick, practical framework that doesn't require building a full model, a small table like the one below can keep your thinking organized without turning investing into a second job.

Lens	*Question it answers*	*Common ways it goes wrong*
NAV (asset value)	What are the properties plausibly worth today, net of debt?	Wrong cap rate assumption; NOI not sustainable; leverage makes small errors huge; ignoring JV structures or preferred claims.
P/FFO or P/AFFO (cash flow multiple)	What am I paying for recurring per-share cash generation vs peers?	FFO/AFFO inflated by adjustments; near-term cash flow peak; leverage and refinancing risks not reflected; bad peer set.

Notice the theme in the "goes wrong" column: valuation errors are rarely math errors. They're assumption errors. The spreadsheet does exactly what you tell it to do, which is why it can be so confident while being so wrong.

Margin of safety: why discounts to NAV are emotionally comforting (and sometimes rational)

Buying below NAV feels safe because it resembles buying a dollar for eighty cents. The phrase "margin of safety" gets used for a reason: if your NAV estimate is roughly right, you have a cushion against small mistakes.

But that margin can evaporate if the NAV itself is sliding. If NOI is weakening, or if cap rates are moving higher, the "dollar" you thought you were buying may be shrinking while you admire the discount. A discount to NAV is only comforting if the assets and cash flows are stable enough that NAV doesn't collapse into your purchase price.

A good discipline is to treat NAV like a range, not a point estimate. Real estate markets are negotiated, not posted. Cap rates move by geography, property quality, tenant profile, and capital availability. If your investment only works at the most optimistic end of the range, it's not a valuation thesis; it's a prayer.

The ending you want: paying a fair price for a durable cash stream

At its best, REIT valuation is not about hunting the absolute cheapest multiple or the deepest discount. It's about paying a price that lets the business do what it does—collect rents, reinvest sensibly, manage leverage—while leaving room for you to earn a solid return without needing perfect luck.

NAV keeps you grounded in the physical world: buildings have market prices, and debt is not optional. Multiples keep you honest about the investment you actually own: a per-share cash engine that can grow, stall, or break depending on leases, capital costs, and management choices.

When both lenses suggest you're paying a reasonable price for a REIT with defendable cash flows, manageable leverage, and a management team that treats capital like it's scarce, valuation stops being a hurdle and becomes a tailwind. That's the quiet advantage: not a flashy bargain, but a price that doesn't require heroics.

3.4 Identifying Quality and Moats

Two REITs can own buildings that look identical on Google Street View and still deliver wildly different outcomes for shareholders. Same city. Same occupancy. Same rent checks arriving every month. One compounds quietly for a decade; the other drifts, dilutes, and eventually "strategically reviews alternatives" (corporate code for *we ran out of good options*). The difference often isn't the real estate. It's the humans and the hidden advantages that make the real estate behave.

Quality in REITs is not a vibe. It's a set of repeatable behaviors and structural edges that show up in the numbers *later*. By the time quality becomes obvious in reported results, the market has usually noticed. The real trick is learning to recognize quality while it still looks like nothing more than boring competence.

Moats in real estate also look different than moats in software. There's no magical codebase. There's no network effect where a million users lock you in. Real estate moats are more like unfair geometry: the asset sits in a location that can't be replicated, under regulations that won't be loosened, with tenants who don't want to move, financed by a balance sheet that can outlast competitors. The best REITs stack these edges until competing starts to feel expensive.

The ``castle'': what makes the properties hard to replace

Most people think real estate competitive advantage is "nice buildings." Nice helps, but it's not the moat. The moat is the *constraint*.

A coastal apartment tower in a supply-constrained neighborhood has a different competitive environment than a garden-style complex where new units can pop up whenever a developer feels optimistic. A last-mile industrial site near a dense population center competes with geometry and traffic patterns, not with marketing. A dominant self-storage operator at the best intersection in town benefits from the uncomfortable truth that people choose storage when they're stressed, busy, or moving—and once their stuff is inside, switching is a chore.

A moat is often just a polite word for "you can't easily build another one of these."

That constraint can come from zoning, scarcity of land, neighborhood politics, environmental restrictions, or the simple fact that the best sites were claimed long ago. When a REIT owns a portfolio of "can't-build-it-again" locations, rent growth becomes less of a negotiation. Tenants can threaten to leave, but they can't easily find an equivalent substitute.

Tenant stickiness: the quiet superpower

The best REIT cash flows aren't just high-they're *stubborn*. They don't disappear easily.

Tenant stickiness is the unglamorous reason. It's the degree to which a tenant resists moving even if you raise rent. Stickiness can come from switching costs (moving a warehouse operation is a headache), operational integration (a medical office tied to a hospital campus), or customer patterns (a grocery-anchored center where shoppers already "know where it is").

Here's the non-obvious part: stickiness turns modest rent increases into a kind of low-drama growth engine. When tenants renew rather than move, you avoid vacancy downtime, leasing commissions, and tenant improvements that eat cash. A sticky tenant base doesn't just support revenue. It protects *AFFO per share,* because it reduces the frictional costs of keeping buildings full.

You can often see stickiness in a REIT's leasing disclosures: renewal spreads, retention rates, and the gap between in-place rents and market rents. But the concept is bigger than the statistics. It's about whether tenants treat the space as a commodity or as infrastructure.

Commodity space invites constant price competition. Infrastructure space invites long-term relationships.

Lease structure: the contract that decides who eats inflation

A REIT's properties might be great, but the lease terms decide who wins the tug-of-war when the world changes.

In some sectors, leases reprice quickly. Hotels can reprice nightly; apartments reprice annually; self-storage can reprice frequently. That can be a gift during inflationary periods, because revenue adjusts upward faster. But it can also be a curse in downturns, because revenue adjusts downward faster too.

Other sectors are built on longer leases—industrial, net lease retail, many office portfolios. Long leases can be stabilizing, but they can also trap the landlord if the lease escalators are too low and inflation is too high. A lease with 1% bumps in a 4% inflation world is a slow leak. A lease with CPI-linked escalators is a different animal.

Lease structure is not "good" or "bad." It's a risk profile. Quality shows up when the lease profile fits the asset and the strategy instead of being an accident of history.

The balance sheet: where good REITs become unkillable

A strong balance sheet is not just defensive. It's offensive.

When credit tightens and weaker landlords are forced to sell assets, refinance at ugly terms, or pause investment, the REIT with liquidity and staggered maturities gets to *choose*. Choice is the rarest luxury in real estate. It lets management buy properties when prices are pressured, fund developments when construction lenders are stingy, and negotiate from strength rather than desperation.

A balance sheet moat often shows up through a few recurring traits:

- Moderate leverage that doesn't require perfect conditions.
- A maturity ladder rather than a near-term "wall" of debt.
- Meaningful fixed-rate debt (or hedged exposure) that prevents sudden interest-expense spikes.
- Unsecured borrowing capacity, which signals lender confidence and increases flexibility.

The deeper point is psychological. A REIT with an overlevered balance sheet is forced to make decisions on the market's schedule. A REIT with a resilient balance sheet makes decisions on *its* schedule. Over long periods, that difference compounds.

This is also where "quality" becomes visible in valuation. The market tends to grant higher multiples and smaller NAV discounts to REITs that can withstand refinancing cycles without turning shareholder distributions into a crisis hotline.

Cost of capital: the compounding advantage that doesn't look like an advantage

In REIT land, the ultimate superpower is a low cost of capital—the ability to raise debt and equity cheaply.

It sounds abstract until you see what it enables. A REIT that can borrow at attractive rates and issue equity at a healthy valuation can buy or develop properties and still grow cash flow *per share*. That last phrase is the key. Growth that doesn't improve per-share metrics is just corporate weightlifting: impressive effort, questionable payoff.

A low cost of capital is not a badge you pin on the company. It's an outcome of other qualities: stable cash flows, transparent reporting, conservative leverage, and a reputation for not doing anything reckless with shareholder money. When those are in place, lenders and equity investors demand less compensation for risk. That cheaper funding becomes a tool to create more growth, which reinforces the reputation, which lowers the cost of capital further.

This is how the best REITs win without shouting. They don't need to outsmart the market every year. They just need to consistently access money at better terms than their competitors.

The ``jockey": management that compounds instead of collecting fees

Real estate attracts capital. Capital attracts managers. Managers attract incentives. Incentives attract behavior. If you ignore that chain, you end up analyzing the buildings while the real risk sits in the boardroom.

Capital allocation is the management skill that matters most. It's the art of deciding when to buy, when to sell, when to develop, when to issue shares, and when to sit on your hands.

The cleanest way to evaluate capital allocation is to look at whether management tends to do the hard-but-smart thing:

- Issuing equity when the stock is expensive relative to NAV, then buying properties at attractive values (good).
- Avoiding equity issuance when the stock is cheap, even if that slows growth (also good).

- Selling assets when private-market values are rich and redeploying into better opportunities (good).

And, of course, the mirror-image behaviors that destroy value:

- Issuing shares when the stock is depressed, diluting existing owners to fund deals that might not even be great (bad).
- Chasing growth for its own sake, especially late in a cycle (bad).
- Treating dividends and buybacks like PR tools rather than capital allocation decisions (bad).

If you want a single sentence that captures management quality: *Do they treat capital like it's scarce, even when markets are generous?* The best teams do. The worst teams act like the ATM will always work.

Internal vs. external management: follow the incentive plumbing

Some REITs are internally managed, meaning the executives work for the REIT itself. Others are externally managed, meaning an outside advisor runs the REIT in exchange for fees that are often tied to assets under management.

External management can create a structural temptation: growth increases fees even if it doesn't increase per-share value. It's not that external management is automatically bad; it's that the incentive plumbing needs to be checked carefully. When the person deciding whether to buy more buildings gets paid more for buying more buildings, you should want unusually strong governance and transparency.

Internal management tends to align incentives better with shareholders-again, not guaranteed, but more naturally aligned. In practice, many investors treat internal management as a positive quality signal, especially when paired with sensible compensation plans that reward per-share performance rather than empire-building.

The point isn't to memorize a rule. The point is to ask: *Who gets richer when the REIT gets bigger? And is "bigger" the same as "better" for current owners?*

Governance as a moat: boring rules that create real money

Good governance rarely makes headlines, which is exactly why it matters. It reduces the odds of catastrophic self-inflicted wounds.

Look for independence and accountability: a board that can say "no," compensation structures that reward long-term per-share results, and disclosure that is consistent and legible. When management communicates in a way that makes comparison easy-clear reconciliations, stable definitions, straightforward explanations-it lowers investor uncertainty. Lower uncertainty can translate into a lower cost of capital. And there it is again: the compounding advantage.

It's easy to underestimate transparency because it feels like etiquette. In markets, etiquette becomes economics.

Disclosure quality: when the footnotes are the product

REITs live and die by trust. They constantly access capital markets, roll debt, issue shares, buy properties, and sometimes sell them. Investors who trust the numbers demand less return for holding the stock. Investors who don't trust the numbers demand a discount, even if the properties are fine.

That's why consistent, comparable disclosure is not just "nice reporting." It's competitive advantage.

A high-quality REIT tends to:

- Provide clear definitions of metrics like FFO and AFFO and stick with them.

- Explain leasing trends with enough detail to separate signal from noise.
- Lay out debt maturities and liquidity in a way that makes risk visible, not hidden.

A lower-quality REIT tends to hide behind glossy presentations, constantly shifting "adjusted" metrics, and selective comparisons. When you feel like you're being managed as a customer rather than treated as an owner, that discomfort is often rational.

Operational discipline: the unsexy difference between ``owned" and ``operated"

Owning buildings sounds passive. Operating them is not.

Quality REITs tend to be excellent operators: they control expenses without starving the properties, they invest in upgrades that earn returns, and they treat tenant relationships like a long-term asset. Poor operators chase occupancy with concessions, defer maintenance, and then act surprised when retention falls. Over time, operations becomes a moat because it's harder to copy than a building address.

This is especially visible in sectors where service matters—apartments, self-storage, hotels, senior housing—where the "product" is not just square footage but experience, responsiveness, and trust.

Operational discipline also shows up in same-store performance. If a REIT can produce steady same-store NOI growth across cycles, it's often a sign that the machine works without constant acquisitions. That kind of internal growth is valuable because it doesn't require issuing shares or taking on incremental leverage.

Scale: not a trophy, a tool

Scale is frequently misunderstood. Bigger is not automatically better. But in real estate, scale can lower costs in ways that are quietly powerful.

Large REITs may borrow at lower spreads, access unsecured debt markets more easily, negotiate better insurance and service contracts, and spread corporate overhead over a larger asset base. Scale can also improve tenant relationships; national tenants like dealing with landlords who can offer multiple locations and consistent execution.

When scale becomes a moat, it's because it creates a measurable advantage: lower cost of capital, lower operating costs, better tenant access, or better data and revenue management. When scale is just size for size's sake, it's a vanity metric wearing a finance costume.

A quality checklist that doesn't read like a checklist

If "quality" feels fuzzy, anchor it to questions that force specific answers:

- Do tenants *need* these locations, or merely *rent* them?
- Would a competitor have a hard time replicating the portfolio in the same markets?
- Can the REIT refinance and invest through a bad credit window without cutting the dividend to the bone?
- Does management repeatedly make per-share accretive decisions, or just growthy ones?
- Is the REIT's disclosure designed to help owners understand risk, or to keep them calm?

Notice how none of those questions can be answered with a single quarter's earnings. That's the point. Moats are slow. They don't announce themselves. They just keep working while other players cycle through strategies, slogans, and "transformational" deals that look good until the bill comes due.

A high-quality REIT is often the one that makes you slightly bored when you read the filings-because the story is consistent, the numbers reconcile cleanly, the risks are acknowledged plainly, and the business doesn't require constant reinvention to keep functioning.

Boring competence, in real estate, is frequently the most expensive thing to buy—because once the market recognizes it, the valuation rarely gives it away for free.

Chapter 4

Navigating Risks and Macro Forces

If a REIT owns great buildings but runs out of cash, the buildings stay standing, but your investment collapses. Real estate feels tangible, but the profits are often dictated by invisible forces: the cost of money, the mood of the market, and the resilience of a balance sheet. In this chapter, we stop looking at the buildings and start looking at the foundation. We'll learn how to survive rising interest rates, spot the difference between a high yield and a trap, and 'stress test' a portfolio so you can sleep well even when the economy is wide awake.

4.1 The Invisible Gravity: Interest Rates and Inflation

The most frightening real-estate headline is rarely about bricks. It's about math. When interest rates rise, nothing about an apartment building's roof changes—but the *price investors are willing to pay for that roof* can change overnight. That's the invisible gravity: rates tug on valuations even when the property is running just fine.

A REIT feels that gravity in two different places, and they're easy to confuse. First is the stock market's immediate mood swing: investors re-price anything "bond-like" (steady cash flow, dividends) when safer yields get more attractive. Second is the slow, operational reality: what the REIT actually earns from rents after expenses and interest. One is a vote-counting machine; the other is a weighing machine. The trouble—and the opportunity—comes from knowing which one you're looking at.

Why higher rates can make perfectly good buildings look cheaper

Real estate is typically valued using a simple relationship: the income a property produces divided by the rate of return investors demand for owning it. In property markets, that demanded return is expressed as a *cap rate* (capitalization rate). If a building produces $1,000,000 of net operating income (NOI) and buyers require a 5% cap rate, the implied value is about $20 million.

$$\text{Value} \approx \frac{\text{NOI}}{\text{Cap Rate}}$$

That formula has the emotional subtlety of a guillotine. If the cap rate moves from 5% to 6%—a change that sounds small enough to ignore—the implied value drops from $20 million to about $16.7 million. The building didn't get worse. The bar for "good enough" got higher.

The cap rate doesn't float in a vacuum. It competes with other ways to earn a return, especially U.S. Treasuries. When Treasury yields rise, investors often demand a higher return from real estate too, because they can suddenly get more income from something that's considered safer and easier to own. Think of it as a spread: *cap rate minus Treasury yield*. That spread is not a law of nature—it widens and narrows with sentiment, growth expectations, and risk appetite—but it's a useful dashboard gauge. When the spread gets pinched, real estate has to "reprice" to restore the appeal.

This is where headlines do their worst work. You'll see "rates up" and "REITs down" and it feels like a verdict on the quality of the properties. More often it's a verdict on discount rates. Higher rates increase the return investors demand; higher demanded returns compress valuations; compressed valuations pull down share prices. It's not personal. It's arithmetic.

Cost of capital: the REIT's internal gravity

Now for the second gravity well: the REIT's cost of capital. A REIT is not just a bundle of buildings; it's also a financing machine. It borrows money, issues shares, buys properties, collects rent, and repeats. When borrowing costs rise, the machine has to work harder to get the same output.

The phrase *cost of capital* can sound like something meant for CFOs and insomnia, but it's just the blended price a REIT pays for its money. If a REIT can borrow at 4% and buy properties at a 6% cap rate, the spread is generous. If it has to borrow at 7% to buy properties at a 6% cap rate, it's trying to drink through a straw with a hole in it.

This spread between property yields (cap rates) and financing costs is one of the cleanest ways to understand why rates matter. REITs grow by acquiring and developing real estate. If the funding becomes expensive, growth slows, deals don't pencil out, and the market stops paying up for expansion stories. In extreme cases, a REIT may even have to sell assets or issue equity at unattractive prices just to keep leverage in check. That's when "higher rates" stops being a valuation story and becomes a survival story.

The market tends to anticipate all of this quickly. Share prices move not when the damage is obvious, but when investors smell it in the air. A REIT that has a debt maturity in three years may see its stock punished today if investors expect refinancing to be painful later. The building's leases don't roll daily, but the stock's expectations do.

The great misunderstanding: rising rates don't automatically doom REITs

Here's a nuance that separates readers from headline-chasers: rising rates are not always bearish for REITs over meaningful stretches of time. Rates can rise for different reasons. If rates rise because inflation is accelerating while growth is shaky, that's nasty—higher financing costs and anxious tenants. If rates rise because the economy is strong, tenants are expanding, and rents are climbing, the REIT's operating engine can offset the financing headwind.

In other words, the question isn't "are rates rising?" It's "what else is rising with them?" Employment? Wage growth? Consumer demand? Corporate profits? Stronger fundamentals can put real wind behind rent growth, occupancy, and rent collections. A rate-hiking cycle that accompanies a healthier economy can still allow REIT cash flows to grow, even as the market applies a harsher discount rate.

That's why REIT investing has this maddening split-screen effect. On one side you have the trading screen reacting to macro expectations. On the other side you have the rent roll reacting to local supply and demand. A patient investor learns to read both without confusing them.

Inflation: friend, foe, and frequent liar

Inflation is often marketed as a tailwind for real estate. The pitch goes: "Prices rise, so rents rise, so real estate wins." Sometimes that's true. Sometimes it's a half-truth with a great suit.

Inflation is simply the general rise in prices over time. Most people see it through consumer measures like CPI, which tracks the cost of a basket of goods and services. But what matters for a REIT isn't the CPI headline by itself. What matters is whether the REIT can *raise rents faster than its costs* and whether its leases allow inflation to show up quickly in revenue.

That last part is the trap. Inflation can be loud in the economy but quiet in a lease contract.

Lease structure is the difference between a thermostat and a painting

A REIT's leases determine how quickly revenue can respond to inflation. Think of lease structure like the difference between a thermostat and a painting. A thermostat responds to the room. A painting stays the same no matter what the temperature is.

Short leases are the thermostat. Hotels re-price nightly. Self-storage can re-price monthly. Apartments typically re-price annually as leases roll. When inflation is pushing wages and prices up, these property types can often adjust rents faster, especially if demand is healthy. That doesn't guarantee higher profits—operating costs can rise too—but it gives the REIT a fighting chance to keep revenue aligned with reality.

Long leases can be the painting. Many net lease and certain office leases run for years. If the lease says the rent increases 1% per year and inflation runs at 4%, the landlord is falling behind in real terms even though the rent is "going up." The cash flow looks stable; the purchasing power quietly leaks.

The really interesting cases sit in between: leases with inflation-linked escalators. Some contracts have fixed annual bumps; others have CPI-based adjustments (sometimes with caps and floors). CPI linkage is not a magic wand—caps can limit upside, and tenant health matters more when expenses jump—but it changes the inflation profile of the cash flows. Two REITs can own properties that look identical in photos and behave completely differently because one has CPI-linked rent resets and the other is locked into sleepy 1% escalators.

Inflation can lift rents and still squeeze margins

Even in property types with fast rent resets, inflation is not a pure gift. A REIT is a business with bills. Labor costs rise. Maintenance and insurance rise. Property taxes can rise. Utilities can rise (sometimes paid by tenants, sometimes by the landlord). Development costs rise, which can make new supply more expensive—good for existing landlords—but it also makes the REIT's own growth projects pricier and riskier.

The key is *operating leverage*. If revenue can re-price quickly and expenses re-price slowly, margins expand. If expenses re-price immediately and revenue is stuck in long leases, margins compress. Inflation is less about the headline number and more about the timing mismatch between revenue and costs.

This is why "real estate is an inflation hedge" is both true and incomplete. Over long periods, real assets often keep up with expected inflation in their capital values and cash flows. But short bursts of inflation—especially surprising inflation—can cause financial markets to demand higher returns right now, which can push valuations down even if the long-run economics are intact. Inflation can be an operating tailwind while simultaneously being a valuation headwind.

That contradiction is not a bug in REITs. It's the core feature. You're holding a long-lived asset financed in a world where discount rates can change quickly.

The refinancing channel: inflation's evil twin

Inflation rarely arrives alone. Central banks often respond by raising interest rates, and that introduces a second-order effect that can overwhelm the first: refinancing.

A REIT with mostly fixed-rate, long-term debt can ride through a rate spike like a ship with a deep keel. The income statement doesn't immediately implode just because the Fed moved. The pain arrives later, when debt matures and has to be refinanced at new rates.

A REIT with a lot of floating-rate debt feels the pain immediately. Floating-rate debt is like a variable-rate mortgage: as benchmark rates rise, interest expense rises. That can chew up cash flow quickly, and since REITs distribute much of their income, they don't usually have the luxury of stockpiling cash for a storm. Floating-rate exposure can turn a macro headline into a direct hit to dividend coverage.

The same inflation print can therefore produce two different stories:

- A REIT with short leases and fixed-rate debt might see revenues adjust upward while interest expense stays stable.
- A REIT with long leases and floating-rate debt might see revenue lag while interest expense jumps.

Both are "real estate." Only one is positioned to enjoy inflation rather than endure it.

REIT prices can drop even when rents are rising

This is the part that feels unfair until you internalize it: REIT shares can fall during periods when the REIT's rent growth is strong. The reason is discounting.

Public markets are constantly re-evaluating what future cash flows are worth *today*. When interest rates rise, the present value of future cash flows generally falls. That's not a real-estate-specific phenomenon; it's finance. A REIT is just a particularly clear case because its cash flows are long-lived and often relatively steady, which makes them sensitive to the discount rate.

So you can have a year where same-store NOI is growing, occupancy is healthy, and management is executing—and the stock is down because the market is repricing the entire sector's valuation multiple. In those moments, the stock market is not grading the REIT's property management skills. It's grading the return requirements of capital.

For investors who only look at price charts, this can look like a business deteriorating. For investors who can separate operations from valuation, it can look like a sale on a business whose cash flows may be improving.

A practical way to read rate risk without becoming a macro tourist

You don't need to predict the next Fed meeting to understand interest-rate sensitivity. You just need a few simple questions that connect macro forces to a specific REIT's reality.

First: *How much debt is fixed versus floating?* Fixed-rate debt buys time. Floating-rate debt buys excitement you probably didn't ask for.

Second: *When does the debt mature?* A REIT with most maturities far out can watch rates rise with less immediate damage. A REIT with a chunk of maturities in the next year or two is effectively holding a ticking coupon book that will be repriced by the market.

Third: *How fast can rents reset?* The lease structure determines whether inflation shows up in revenue quickly or slowly. The faster rents can adjust, the more likely operating cash flow can keep up with an inflationary environment—assuming demand isn't collapsing.

Fourth: *Who pays the expenses?* Some leases push property-level expenses to tenants (common in net lease structures), while others leave more of the inflation burden with the landlord. Expense pass-through provisions can be the difference between inflation boosting NOI and inflation merely boosting receipts while costs eat the gain.

Notice what's missing: guessing where the 10-year Treasury will be next quarter. Rate forecasting is a hobby. Balance-sheet and lease analysis is an edge.

The spread that quietly governs everything

There's a simple mental model that helps tie rates, cap rates, and REIT valuations together: investors compare the yield on real estate to the yield on safer assets. When Treasuries yield very little, a 5% property cap rate looks delicious. When Treasuries yield more, that same 5% looks less compelling unless rent growth is strong or the cap rate moves higher (meaning the property price moves lower).

This is why real estate valuations can fall even if the economy is steady: the alternative options improved. Capital is competitive. It doesn't hate real estate. It just demands to be paid.

For REITs, this shows up as changes in *NAV sentiment* (net asset value). If private-market cap rates move up, the estimated value of the REIT's property portfolio may be marked down. Public REIT prices can trade at a premium or discount to that estimated NAV, and those discounts can widen dramatically when investors fear that private-market prices haven't fully adjusted yet. In rate shocks, public markets often move first and argue later.

That gap between public prices and private marks is not just drama—it can influence corporate behavior. A REIT trading at a deep discount to NAV may find it unattractive to issue new equity for acquisitions, slowing growth. If the discount is severe and liquidity is needed, management may have to sell assets, cut development, or focus on balance-sheet defense. Rates and spreads aren't abstract; they shape what management can do.

Inflation protection is not a slogan; it's a contract

If someone tells you a REIT is "inflation-protected," ask: *By what mechanism?* Not by vibes. Not by the age of the buildings. By leases.

A CPI-linked escalator with a reasonable cap might provide partial inflation pass-through, especially when inflation is moderate and sustained. Short leases in high-demand markets can allow landlords to reset rents at market. Development constraints and replacement cost inflation can limit new supply, indirectly supporting rent growth for existing properties. These are concrete channels.

But there are also channels that work the other way. Long leases with weak escalators can trap cash flows. Rent control or regulatory restrictions can limit pass-through. Tenant health can deteriorate if inflation squeezes consumers or operating margins, increasing vacancy risk or credit losses. And higher rates used to fight inflation can pull down valuations. Inflation is a many-headed creature; you don't get to fight only the head you like.

So the right framing is not "REITs win in inflation." The right framing is: *some REIT cash flows can adjust with inflation, depending on lease duration, rent escalators, expense pass-throughs, and tenant strength—but valuations can still fall if discount rates rise faster than cash flows.*

That sounds less sexy than a slogan. It's also more likely to keep you solvent.

The investor's advantage: time arbitrage

The stock market reacts to rate expectations at the speed of a tweet. Buildings react to inflation at the speed of a lease rollover. That mismatch creates a strange advantage for long-term REIT investors: you can sometimes buy cash flows that will adjust over time at prices that assume they never will.

When a rate shock hits, the market can punish every REIT indiscriminately. In the chaos, a storage REIT with month-to-month pricing power can trade down alongside a long-lease REIT whose rents are locked for years. The difference is that one business can potentially reprice its revenue quickly while the other needs time—and in markets, time is often mispriced.

This is not a promise that every drawdown is a bargain. It's a reminder that macro-driven selling can flatten important distinctions. If you can keep your curiosity alive when everyone else is panicking, you can look for REITs with (1) manageable refinancing risk and (2) genuine inflation-responsive revenue. Those two together are rare enough to matter.

A final, grounding thought: you're not buying ``real estate," you're buying a set of sensitivities

A REIT is a portfolio of properties, yes. But it's also a portfolio of *sensitivity settings*:

- sensitivity to discount rates (valuation),
- sensitivity to borrowing costs (cash flow),
- sensitivity to rent resets (inflation pass-through),
- sensitivity to tenant health (credit),
- sensitivity to expense inflation (margin).

The invisible gravity of rates and inflation doesn't make REIT investing fragile. It makes it *legible*. The forces are big, but the channels are knowable. When you learn to read the lease and the balance sheet as carefully as you read the dividend yield, macro headlines stop feeling like thunder. They start feeling like weather—sometimes dangerous, often noisy, always worth respecting, and rarely the whole story.

4.2 Fortress or House of Cards? Analyzing the Balance Sheet

A REIT can own gorgeous buildings and still be a bad investment. That sounds rude, but debt doesn't care about curb appeal. Lenders don't accept "great location" as payment. They accept cash, on schedule, in the exact amounts promised—and the balance sheet is the document that tells you whether those promises are comfortable...or clenched-teeth optimistic.

The seductive mistake is to treat a REIT's debt like background noise. It isn't. Debt is the amplifier knob on every outcome. In good times it boosts returns; in bad times it turns an ordinary problem (a few vacancies, a temporary rent dip, a higher insurance bill) into a capital-markets emergency. The properties produce the income, but the *financing terms decide who gets to keep it.*

A quick translation: what the balance sheet is really saying

A REIT balance sheet is a snapshot of ownership and obligations. The properties and other assets sit on one side; the funding sources sit on the other. Equity is what's left after subtracting liabilities from assets. That part is straightforward. The interesting part is that real estate is typically financed with a mix of long-lived assets and rolling promises: debt that comes due, gets refinanced, and comes due again.

If you want a single mental image, picture the REIT as a landlord with multiple mortgages, each with its own rate and maturity date, plus a credit card (the revolver), plus a set of rules (covenants) written by a very humorless person. A "fortress" REIT is the one whose rules are loose, whose maturities are spread out, whose cash cushion is real, and whose interest expense doesn't jump every time the central bank clears its throat.

Leverage: the most misunderstood word in REIT land

Leverage is just borrowing. The reason it gets mystical is that different people measure it differently, and REITs are allergic to one-size-fits-all ratios. Some investors look at debt-to-assets. Others prefer debt-to-EBITDA (or net debt-to-EBITDA). In REIT reporting, you'll also see debt-to-gross assets, debt-to-total market capitalization, and various "look-through" versions that adjust for joint ventures.

Rather than memorize every acronym, start with the purpose of the ratio: *how much trouble does the REIT get into if cash flow dips?* Leverage matters because it fixes a chunk of the cost structure. Rent can go down. Interest payments do not politely adjust.

Debt-to-EBITDA is popular because it ties debt to the earnings power that services it. EBITDA is not perfect for real estate, but it's close enough to be useful as a stress thermometer: higher leverage means the REIT needs more consistent cash flow to avoid awkward conversations with lenders and rating agencies.

Debt-to-assets (or debt-to-gross assets) is a different lens: it asks what portion of the property portfolio is financed with debt. It's a sturdier measure when earnings are temporarily distorted, but it relies on asset values that can be slow to adjust in reports. In fast-moving markets, "assets" can be a polite fiction for a while.

The practical investor takeaway is simple: leverage is not only about the number; it's about the *distance to the cliff*. Two REITs can both have 40% debt-to-assets, but one might have predictable, diversified rent checks and the other might be riding a single tenant and a prayer. Leverage is the spark; business quality is the gasoline.

Secured vs. unsecured: who holds the keys?

Debt comes in two broad flavors. *Secured* debt is tied to specific properties. If the REIT can't pay, the lender has a direct claim on those assets. *Unsecured* debt is backed by the REIT as a whole, not by a particular building.

Unsecured debt is often a sign of a more mature, institutionally trusted balance sheet. Lenders are essentially saying, "We'll lend to you because your overall enterprise is strong." Secured debt can be perfectly reasonable—especially for specialized assets or development—but too much secured financing can become a trap. It can limit flexibility, restrict asset sales, and make refinancing harder because more and more of the portfolio is already pledged.

If you're trying to decide whether a REIT is building a fortress or stacking cards, ask: *If they needed to raise cash quickly, could they sell an asset without negotiating with three lenders and an army of lawyers?* Flexibility is a financial superpower when markets freeze.

Fixed vs. floating: the difference between annoyance and panic

Interest rates don't hit every REIT the same way. The clearest "tell" is the fixed-versus-floating mix.

Fixed-rate debt is boring, which is a compliment. The interest expense is largely known. The pain of higher rates arrives later, when the debt matures and must be refinanced.

Floating-rate debt is a direct pipeline from the rate market to the income statement. If benchmark rates rise, interest expense rises. This is where an investor who likes dividends should become intensely curious. A REIT can have a stable-looking portfolio and still see distributable cash squeezed simply because the interest bill inflated.

Floating-rate debt isn't inherently evil. It can be a tool for short-term needs, bridge financing, or opportunistic flexibility. It becomes dangerous when it's large enough that a rate spike forces management into defensive moves: cutting the dividend, issuing equity at depressed prices, or selling assets at the wrong time. In a calm market, floating rate looks cheaper. In a stressed market, it looks like a hidden variable you didn't sign up for.

The debt maturity ladder: where the balance sheet tells the truth

Leverage ratios are like blood pressure: vital information, but they don't tell you when the next heart attack might happen. For that, you need the debt maturity schedule—often called the *debt ladder*.

A debt ladder lists how much debt comes due each year. The best ladders look boring: maturities spread out over many years, with no single year that demands heroics. The worst ladders look like a cliff: an enormous amount due soon, followed by a long quiet stretch that won't matter if the REIT can't survive the near-term refinancing.

Refinancing risk is rarely about today's interest expense. It's about *tomorrow's access to capital*. When a big maturity hits, the REIT must either:

- refinance (issue new debt to repay old debt),
- repay with cash (often unrealistic at scale),
- sell assets,

- or issue equity.

In friendly credit markets, refinancing can be routine. In hostile markets, it can be punitive or impossible. A "maturity wall" turns capital markets into a landlord's most important tenant: if the market doesn't renew the lease, you have a vacancy measured in billions.

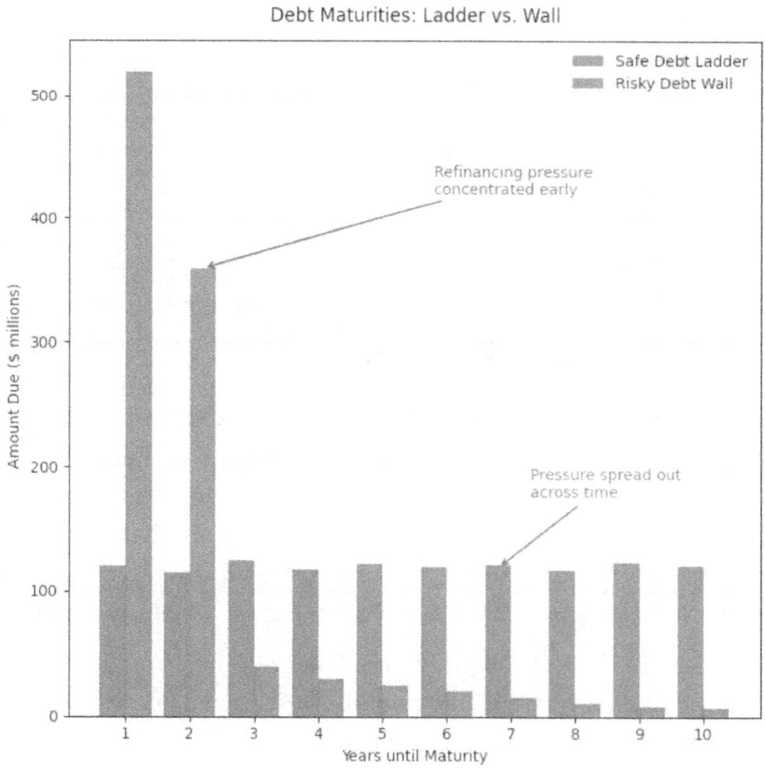

A safe ladder does something psychologically comforting: it turns a terrifying, binary event ("we must refinance $2 billion next year") into a series of manageable chores. It also gives management options. If credit spreads widen in one year, they can wait for a better window. If the equity market is receptive, they can issue a little. If asset values are strong, they can sell selectively. A wall removes optionality. A wall forces a decision under pressure, which is exactly when you get the worst prices.

This is why debt maturity is the closest thing to a "ticking clock" you'll find in a REIT. Buildings can hold value for decades. A maturity date is a deadline.

Liquidity: the cash oxygen mask

Liquidity is what keeps a REIT from making stupid decisions in a stressful moment. It's the ability to meet obligations without selling crown-jewel assets at clearance prices. Liquidity usually comes from three places:

- cash on the balance sheet,
- an undrawn (or partially undrawn) revolving credit facility,
- and predictable free cash flow.

Cash is obvious. Revolver availability is less obvious but arguably more important: it's the corporate equivalent of having a line of credit for emergencies. The catch is that revolvers come with covenants and conditions, and in a true crisis, banks can get cautious. A revolver is not a magic portal to infinite money; it's a contract that assumes you remain a reasonably healthy borrower.

A well-run REIT treats liquidity like fire insurance. You don't buy it because you enjoy paying the premium. You buy it because you cannot predict which random Tuesday the world decides to misbehave.

Covenants: the fine print that can move faster than the rent roll

Debt covenants are rules lenders impose to protect themselves. They might limit leverage, require a minimum interest coverage ratio, or restrict secured borrowing. Covenants are the reason two REITs with similar assets can have very different risk profiles: one has room to breathe, the other has a collar around its neck.

The sneaky part is that covenants can tighten *without management doing anything wrong*. If property values fall, leverage ratios can look worse. If EBITDA dips due to vacancy or tenant issues, coverage ratios can deteriorate. If interest rates rise and floating-rate debt becomes more expensive, coverage can compress again. Covenants are sensitive to the same macro forces that hit valuations and cash flows—which is why they tend to matter most when you least want them to.

A fortress balance sheet isn't just low-ish leverage. It's leverage that is low enough to remain compliant under stress. The goal is not to look great in perfect weather. The goal is to stay functional in bad weather.

Interest coverage: the simplest test of comfort

If you only ran one quick check on a REIT's debt burden, it would be interest coverage: how many times operating earnings cover interest expense. The higher the better, all else equal. Think of it as financial slack.

Coverage ratios can lull you into complacency if you ignore the debt structure. A REIT with decent coverage today but a large floating-rate exposure can see that coverage shrink quickly as rates rise. A REIT with good coverage but a near-term maturity wall can face a huge step-up in interest expense at refinancing. Coverage is a snapshot; the debt ladder is the movie.

So treat interest coverage like a smoke alarm. It's not the fire, but it tells you whether you should start looking for it.

Preferred equity and hybrids: the "almost debt" that behaves like debt when you're stressed

REIT capital structures sometimes include preferred stock or hybrid securities. Preferred dividends are not technically interest, and skipping them may not trigger immediate default the way missed interest payments do. But in practice, preferred obligations behave like fixed charges. They're expected to be paid, and markets interpret non-payment as a distress flare.

When evaluating leverage and fixed-charge coverage, mentally group preferred dividends with interest expense. They're both claims that come before the common shareholder gets anything. A fortress structure is one where the common dividend is supported by healthy residual cash flow after these fixed claims—not merely by optimism.

The equity issuance test: can they raise money without humiliating you?

One of the strangest realities of REITs is that issuing equity is not automatically a negative. For many REITs, equity issuance is a normal funding tool. The question is *price*.

A REIT trading at a premium to the value of its assets (or simply at a high multiple relative to cash flow) can raise equity efficiently and buy properties accretively. A REIT trading at a deep discount may find that issuing equity punishes existing shareholders—it's like selling slices of pizza at half price to pay the rent.

This is where balance-sheet quality loops back into returns. A fortress REIT often maintains enough financial credibility that it can access capital markets on reasonable terms. A fragile REIT gets trapped: the stock price falls, equity becomes expensive, debt becomes expensive, and suddenly growth plans turn into triage.

You don't need to predict whether a REIT will issue equity. You just want to own the kind that *can* if it must, without doing it at desperation prices.

A "fortress" checklist that doesn't read like a spreadsheet

A fortress balance sheet has a certain vibe, and it's not subtle. Management talks about maturities the way responsible adults talk about retirement savings: early, clearly, with receipts. The debt ladder is smooth. Floating-rate exposure is limited or hedged. Liquidity is ample. Covenants are described in a way that sounds like they've actually been stress-tested, not merely copy-pasted.

A house-of-cards balance sheet has its own tells. The company is always "evaluating options." It has a big maturity soon and an oddly relaxed tone about it. It leans heavily on "non-core asset sales" that never quite happen. It boasts about a dividend while quietly refinancing at worse and worse terms. It speaks in adjectives when you want numbers.

The point isn't to develop cynicism. The point is to develop pattern recognition. Real estate is slow. Debt deadlines are fast. The best REITs are built so that slow problems stay slow.

How to read the disclosures like an adult, not a tourist

Most of what you need is not hidden. It's just buried in places investors skip because they're less exciting than property photos.

Look for the debt maturity schedule in the supplemental package or notes. Find the fixed-versus-floating breakdown. Check the weighted-average interest rate on debt and the weighted-average maturity. Scan for comments about hedges (interest rate swaps and caps) if floating exposure exists. Identify liquidity: cash plus revolver capacity, and any near-term obligations that consume it (debt maturities, development commitments, preferred dividends).

Then do one piece of "napkin math" that changes the whole feel of the risk: imagine the next refinancing happens at a meaningfully higher rate than today. If the REIT refinances $1 billion and the rate is 3 percentage points higher, that's $30 million of additional annual interest expense. That $30 million comes from somewhere. If the REIT's cash flow easily absorbs it, you're probably looking at a fortress. If it would eat the dividend, you're looking at a balance sheet that only works in a particular interest-rate climate.

The cleanest truth about REIT risk

A REIT doesn't usually die because its buildings suddenly become worthless. It dies because it runs out of time. Refinancing risk is time risk. Liquidity is time insurance. A debt ladder is a time management system.

When you own the fortress, volatility feels like weather. When you own the house of cards, volatility feels like a structural inspection. The market can stay irrational longer than you can stay funded—and the balance sheet is the place where that old saying stops being a quote and becomes a diagnosis.

4.3 The Enemy Within: Behavioral Biases and Yield Traps

The most dangerous thing in REIT investing is not a leaky roof or a vacant building. It's the little voice that whispers, "Look at that yield." Income investors like to think of themselves as disciplined—the sensible adults of the market, collecting rent checks while everyone else chases shiny growth stocks. And then a REIT flashes a double-digit dividend yield like a neon sign in a dark alley, and discipline suddenly develops a blind spot.

A high yield can be a gift. It can also be a distress flare. The problem is that both look identical at first glance: a percentage next to the word *yield*. The market does not label one "bargain" and the other "trap." It just prints the same font and lets you decide whether you're investing or volunteering.

Why yield is so persuasive (even when it shouldn't be)

Yield scratches a very specific itch. It feels concrete. Price charts wiggle. Headlines contradict themselves. But a dividend yield feels like a paycheck you can plan around. In a world where financial advice often boils down to "be patient," yield offers immediacy: you don't have to be right *eventually*; you get paid *now*.

That psychological comfort is exactly why yield can be weaponized by your own brain. A big yield number can make you relax your standards without noticing. You start thinking, "Even if the price goes nowhere, I'm getting 12%." But the market is rarely handing out 12% with a smile and no strings. More often it's saying, "We think this payout is in danger," and the yield is high because the *price collapsed*, not because the business is suddenly generous.

Here's the mechanical truth that turns "income opportunity" into "yield trap" in one sentence: *Dividend yield is a fraction, and the denominator can fall fast.* If a REIT pays $2 per share and the stock drops from $40 to $20, the yield doubles from 5% to 10% without the company doing anything heroic. In fact, that yield spike can happen precisely because investors are anticipating something unheroic: declining cash flow, refinancing stress, a tenant failure, or a dividend cut.

Yield is not a reward label. It's a temperature reading. Sometimes the temperature is high because it's summer. Sometimes it's high because the house is on fire.

The yield trap: a bargain costume over a deteriorating business

A yield trap is a stock that looks cheap because its price has fallen, making the trailing dividend yield look juicy—but the underlying economics are getting worse, and the dividend is likely to be reduced. The trap works because it frames the decision incorrectly. You think you're choosing between "high yield" and "low yield." You're actually choosing between "sustainable payout" and "future disappointment."

The disappointment usually arrives in a familiar sequence. First, the stock falls as informed investors exit. Second, the yield rises because price fell. Third, new buyers rush in, attracted to the now-eye-popping yield. Fourth, reality catches up: management cuts the dividend to protect the balance sheet or because cash flow no longer supports the payout. Fifth, the stock falls again because the investor base that owned it *for income* sells the moment the income is reduced.

If that sounds dramatic, it's because it is. Yield traps don't merely underperform; they can create a double injury: a capital loss plus an income loss. They are the investing equivalent of buying a "discount" parachute.

The sneaky bias: recency turns today into forever

One of the most common investing mistakes is assuming the recent past is a reliable map of the future. If a REIT has paid a steady dividend for five or ten years, it's emotionally tempting to treat that dividend as part of the company's identity—like it's printed on the logo. But dividends are not personality traits. They're financial decisions.

Recency bias shows up in income investing as a kind of lazy extrapolation: "They've paid it, so they'll keep paying it." That's sometimes true—for high-quality REITs with conservative balance sheets and resilient cash flows. But the same mental shortcut can lead you to ignore obvious changes in the environment: interest costs rising, debt maturities approaching, tenants weakening, or property values compressing. The dividend doesn't break because the past was bad. It breaks because the future is different.

The market is often forward-looking in price, but investors can be stubbornly backward-looking in narrative. A yield trap is what happens when you buy the narrative and ignore the math.

Confirmation bias: how investors build a case like a defense attorney

Confirmation bias is the tendency to seek information that supports what you already want to believe and downplay the rest. It's not just "being optimistic." It's *constructing a filtered reality*.

With REITs, confirmation bias has a particularly charming costume: the "income thesis." Once you decide you like a yield, you start collecting supportive facts. "They own good properties." "Real estate always comes back." "The dividend has been paid for years." "Management said they're committed to shareholders." Each statement may be partly true, but together they can form a fortress made of slogans.

Meanwhile, the opposing evidence—the stuff that belongs in the front row—gets demoted. "Yes, but the debt is mostly floating." "Yes, but a huge chunk matures next year." "Yes, but rent escalators are tiny and expenses are rising." "Yes, but the tenant is shaky." Confirmation bias doesn't require you to lie to yourself. It only requires you to *rearrange the importance* of what you already know.

If you want a simple antidote, ask a question that forces the brain to argue the other side: *What would have to be true for this dividend to be cut?* If you can't produce a plausible answer, you're not analyzing; you're auditioning for a fan club.

Familiarity bias: the comfort of what you can picture

Familiarity bias is preferring what feels known. In REITs, it can show up as over-weighting the property types you personally understand: apartments because you rent one, shopping centers because you shop there, warehouses because you've seen delivery trucks. There's nothing wrong with investing in what you can explain. The danger comes when "I recognize it" becomes "it must be safe."

A net lease REIT with tenants you've heard of can feel safer than a data-center REIT whose customers are hyperscalers you can't name. But tenant brand recognition is not the same as tenant credit strength, lease terms, or balance-sheet resilience. Familiarity can also push investors into concentrated bets: buying a handful of REITs that all feel "safe" because they're easy to imagine, even if they share the same macro vulnerabilities.

The market loves to punish unrecognized risks. Familiarity bias is how you make those risks invisible until the punishment arrives.

Noise trading: when the dividend becomes entertainment

REITs produce a steady stream of small news: quarterly earnings, occupancy updates, cap-rate chatter, property sales, refinancing announcements, guidance tweaks. If you watch all of it too closely, investing turns into a reactive hobby. You start trading the dividend like a sports score: up, down, did they "beat," did they "miss."

Noise trading is not the same as being informed. It's being *stimulated*. The more you react to every data point, the more you risk selling high-quality REITs at exactly the wrong time—when headlines are loud but fundamentals are intact—and buying riskier ones because the yield looks irresistible right after a price drop.

There's a cruel irony here: the investors most attracted to stable income can end up with unstable outcomes because they turn stability into something they constantly monitor and interfere with. A watched pot does not boil faster. It just makes you hungry and impatient.

Reaching for yield: when ``income" becomes a risk disguise

"Reaching for yield" is the polite institutional term for a very human behavior: accepting more risk than you normally would because safe yields feel inadequate. When cash and bonds yield little (or when inflation makes "safe" yields feel like a loss), investors start scanning for alternatives that promise more income. The danger is that the additional yield often comes from hidden fragility: leverage, weak tenants, thin margins, or a balance sheet that only works if refinancing stays easy.

In REITs, reaching for yield frequently shows up as a preference for the highest trailing yields in the sector, as if yield is a score. But the highest yields are often attached to the least trusted cash flows.

You're not being paid extra because the market is kind. You're being paid extra because the market is nervous.

A useful way to think about it: yield is like the interest rate on a personal loan. When the rate is very high, it's not because the lender wants to spoil you. It's because they're pricing the chance they won't get their money back.

The seductive arithmetic error: confusing yield with return

Yield is one component of return. It is not the whole meal. Total return is dividend income plus price change. In a yield trap, the dividend can be high while the price erodes, leaving you with a mediocre or negative total return. Worse, when the dividend gets cut, the "income advantage" that justified the risk vanishes, and you're left holding a bruised stock you bought for a paycheck that no longer exists.

The psychological trick is that dividends feel like a gain even when they're not. If you receive $1 in dividends and the stock drops $1, you haven't made money. You've just converted part of your investment value into cash. That conversion can be useful, but it's not alchemy.

A REIT dividend is best thought of as a *distribution of cash flow*, not a guarantee of profit. When the market is pricing a REIT at a distressed level, the dividend may be partly returning your own capital to you in slow motion.

A practical red-flag checklist for ``too-good-to-be-true" yields

Any time a REIT's yield is more than *roughly* twice what you see from the broader, boring parts of the REIT market, treat it as a smoke alarm. It might be a false alarm. But you don't remove the batteries just because you like quiet.

Here are the red flags that deserve a hard stare *before* you buy:

1) The payout is covered by a fragile number

REITs are often evaluated using FFO (funds from operations) or AFFO (adjusted funds from operations), which are cash-flow-like metrics meant to reflect the money available to pay dividends. If the dividend consumes nearly all AFFO—or worse, exceeds it—the margin for error is thin. Thin margins and leverage are a dangerous pairing.

Watch for creative "add-backs" that make coverage look healthier than it feels. If the story relies on a special definition of cash flow that only management uses, be suspicious. Dividends are paid with actual cash, not customized metrics.

2) The debt clock is loud

A high yield paired with near-term maturities is an especially sharp warning. If a REIT must refinance a large amount soon, its future interest expense is uncertain, and that uncertainty belongs in your dividend expectations. A high yield can be the market's way of saying, "We think this will get repriced."

The deadliest version is the combination of a high yield, a maturity wall, and a weak stock price. That's when refinancing pain can force equity issuance at unattractive prices or asset sales that reduce earnings power.

3) The balance sheet has ``variable-rate surprise" built in

Floating-rate debt can turn a macro move into an immediate cash-flow squeeze. If a REIT is paying a large dividend and also carrying substantial floating-rate exposure, you're effectively betting that rates won't stay high or won't rise further. That's not necessarily a bad bet, but it's a bet—and many investors make it accidentally.

If the REIT says it's hedged, look for the *extent and duration* of the hedges. A hedge that expires next year is not a permanent shield; it's a timed umbrella.

4) The properties are fine, but the tenants are not

A property can be beautiful and still produce unreliable rent if the tenants are financially weak. High yields sometimes cluster in REITs exposed to troubled industries, lower-quality retailers, or single-tenant situations where one failure matters a lot. Tenant concentration is a quiet risk that becomes very loud in a downturn.

If you see heavy reliance on a few tenants, ask what happens if one leaves or renegotiates. A REIT with diversified tenants can absorb a punch. A concentrated one can break a rib.

5) The ``dividend story" is doing too much work

When the primary argument for owning the REIT is the dividend itself, you're on dangerous ground. A strong REIT can be described in multiple ways: property quality, rent growth, lease terms, balance sheet, management discipline. A fragile REIT often gets described in one way: "the yield."

If the investment thesis collapses the moment you remove the dividend, you're not investing in a business. You're renting a headline.

Non-traded REITs: the yield trap can come with a lock on the door

Yield-chasing has an extra twist in non-traded REITs (those not listed on a major exchange). Because they don't trade daily, they can appear calmer. The account statement may show a stable value, the distribution may arrive on schedule, and the whole thing can feel like a high-yield savings account with better marketing.

But illiquidity can hide risk rather than eliminate it. If you can't easily sell, you can't easily discover the true market price of the investment. Distributions can be funded in ways that look smooth while the underlying economics are bumpy. Fees and selling commissions can be substantial. Redemption programs can have limits or be suspended at the worst time.

The psychological danger is that investors treat the distribution as proof of safety. It isn't. It's just a payment. The right question is the same as always: *What is the sustainable cash flow after all expenses, debt costs, and necessary reinvestment?* If you can't answer that clearly, the yield is not your friend. It's a distraction.

A better obsession than yield: durability

Income investing works best when you obsess over the boring inputs that create durable cash flow. That means balance-sheet resilience, sensible payout ratios, well-structured leases, and tenants who can pay rent in ugly economies. It means accepting that an 5–7% yield that grows is often more powerful than a 12% yield that gets cut.

The market's cruelest joke is that the highest yields often belong to the companies least able to keep paying them. The kindest lesson is that you don't have to take that bait. You can demand a dividend that is earned, not advertised.

A final gut-check that helps: before buying a high-yield REIT, imagine the dividend is cut by 30% tomorrow. Would you still be happy owning the underlying business at today's price? If the honest answer is no, the yield isn't income. It's a lure.

4.4 Stress-Testing: Playing Financial Wargames

The market has a favorite prank: it waits until you feel smart, then tests whether you're merely optimistic. Stress-testing is how you flip the script. It's not about predicting the next recession or guessing where interest rates will land. It's about asking a simpler, more useful question: *If something ugly happens, do I still get to keep playing?*

That's the real goal. Not perfection. Not bragging rights. Survival with dignity. Because in REIT investing, the catastrophic outcomes tend to come from a short list of repeat offenders: a sudden cash-flow hit, a refinancing squeeze, or a tenant problem that spreads faster than your confidence.

The good news is that you don't need fancy software, a PhD, or access to a bank's risk department. You need honesty, a pencil, and the willingness to temporarily ruin your own day.

What stress-testing is (and what it absolutely isn't)

A stress test is a *hypothetical* scenario designed to see how a business or portfolio behaves under pressure. It's like checking the fire exits before the smoke shows up. The point is not to be "right." The point is to discover where the weak beams are while you still have time to reinforce them.

The most common misuse of stress-testing is turning it into a forecast. Investors run a scenario, decide it looks bad, and then assume the bad thing *will* happen. That's not stress-testing; that's just anxiety with extra steps. A well-built stress test is severe-but-plausible. It's uncomfortable enough to expose fragility, but not so apocalyptic that everything fails and nothing is learned.

Another common misuse is treating stress-testing like a single number, a pass/fail score. Real stress-testing is more like a conversation with your portfolio. You're listening for where it starts to wheeze: the REIT with the maturity wall, the one with the single mega-tenant, the one whose dividend coverage is fine until it isn't.

The only two outputs that matter: cash and choices

Stress scenarios can get elaborate, but the results tend to funnel into two practical questions.

First: *What happens to cash flow?* In REIT terms, that usually means what happens to AFFO (adjusted funds from operations) or whatever cash-flow proxy you use to judge dividend coverage.

Second: *What choices does management have?* Cash flow can dip and still be survivable if the balance sheet gives management options: liquidity, staggered maturities, unencumbered assets, the ability to issue equity at a reasonable price, or at least enough slack to wait for conditions to improve.

If cash falls and choices disappear at the same time, that's when "temporary stress" becomes "permanent impairment." Your job as an investor is to avoid portfolios that rely on perfect weather.

Set up your napkin math (without building a spreadsheet empire)

Start with three numbers for each REIT you own (or want to own):

- Annual AFFO (or AFFO per share times shares outstanding)
- Annual dividend payout
- Net debt and the rough mix of fixed vs. floating debt, plus near-term maturities

You don't need perfection. You need order-of-magnitude clarity. If your estimate is off by 5%, that's fine. Stress-testing is about *direction and sensitivity*. A REIT that breaks in your rough model is unlikely to become unbreakable in reality.

Then pick your scenarios. Keep them few and sharp. Too many scenarios is just procrastination in a lab coat.

Scenario 1: The Recession (vacancies rise and rent growth gets humbled)

Recessions don't announce themselves by politely lowering your dividend. They show up as tenants making odd phone calls. "We'd like to renegotiate." "We're consolidating locations." "We're delaying expansion." The building doesn't vanish. The rent check does.

A simple recession stress for many property types looks like this:

- Occupancy drops by 5–10 percentage points (or vacancy rises by that amount)
- New leases are signed at lower rents, or rent growth stalls
- Bad debt expense rises (more tenants pay late or not at all)

To translate that into napkin math, focus on NOI (net operating income), the property-level cash flow before corporate costs and interest. If you don't have NOI handy, you can approximate the impact on AFFO directly with a haircut assumption.

Example: Suppose a REIT generates $500 million of annual AFFO and pays $350 million in dividends. That's a comfortable cushion. Now impose a recession haircut: AFFO down 15%. AFFO becomes $425 million. Dividend coverage tightens, but the dividend still fits. That's the kind of stress you can live with.

Now try the same with a REIT that generates $500 million of AFFO but pays $480 million in dividends. A 15% hit takes AFFO to $425 million while the dividend remains $480 million. The math is now yelling at you. Management can either fund the dividend with asset sales/debt (not a long-term strategy), cut the dividend, or hope the recession ends quickly. In stress-testing, hope is not a plan; it's a confession.

The recession scenario is not just about the dividend. It's about *knock-on effects*. Lower NOI can raise leverage ratios. Higher leverage can trip covenants. Covenant pressure can restrict access to credit. Restricted access makes refinancing harder. Suddenly the recession is not merely an earnings story; it becomes a balance-sheet story. That's how stress compounds.

Scenario 2: The Rate Spike (interest expense jumps where you're most exposed)

Rate risk isn't only "rates up." It's "rates up *and* spreads wider *and* lenders become picky." In real stress, you don't get a clean, academic shift in the yield curve. You get messy financing conditions: floating-rate costs jump quickly, refinancing is priced harshly, and the market starts demanding more compensation for lending to anything with leverage.

A simple rate-spike stress can be modeled two ways:

A) Floating-rate shock

Assume interest expense doubles on floating-rate debt. That sounds extreme until you remember that floating-rate costs can move fast in tightening cycles, especially if the REIT's credit spread widens too.

Napkin math approach: Identify how much debt is floating and estimate the extra interest.

$$\Delta\text{Interest} \approx \text{Floating Debt} \times \Delta\text{Rate}$$

If a REIT has $2 billion of floating-rate debt and the effective rate increases by 3 percentage points, that's $60 million less cash per year. That $60 million comes directly out of AFFO.

If the dividend cushion was $80 million, you're now one bad quarter away from a cut. If the cushion was $300 million, it's unpleasant but manageable.

B) Refinancing shock (the quiet killer)

Even a REIT with mostly fixed-rate debt faces refinancing risk. The stress here is a "maturity wall year" where the REIT must refinance a large chunk at higher rates.

Model it like this: pick the amount maturing in the next 24 months and assume the new rate is meaningfully higher than the old one (say +200 to +400 basis points, depending on how nasty you want the scenario to be). Then compute the added annual interest.

This scenario is where debt ladders earn their keep. A REIT with staggered maturities might only have to refinance a modest slice each year; a REIT with a wall can be forced into a full-contact fight with the bond market at the worst time. Stress-testing turns that abstract schedule into a very personal question: *Can they refinance without sacrificing the dividend or issuing equity at an ugly price?*

Also note the curve-shape issue. Sometimes short rates rise more than long rates (a painful environment for floating debt). Sometimes long rates rise and cap rates follow (a painful environment for asset values and NAV). A serious stress test doesn't assume the world moves in one clean line. It asks, "Where am I fragile: cash flow today, or valuation and refinancing tomorrow?"

Scenario 3: The Black Swan (lose a top tenant, and watch how fast the story changes)

Black swans aren't always global crises. Sometimes they're embarrassingly specific: your largest tenant goes bankrupt, a major operator decides to consolidate, a hospital system changes strategy, a retailer gets disrupted, a corporate client stops renewing. The portfolio doesn't need to be "bad" for this to hurt. It only needs to be *concentrated.*

A clean black swan stress looks like this:

- Assume the top tenant (or top 3 tenants) stops paying rent
- Assume it takes 12–24 months to re-lease the space
- Assume re-leasing requires capital spending (tenant improvements, leasing commissions)

This is where investors learn the difference between a high-quality building and a high-quality lease. The building may be desirable, but re-leasing takes time and money. During that time, the REIT's cash flow is hit and its leverage rises. If the REIT already had tight dividend coverage or near-term maturities, one tenant event can start a chain reaction.

Napkin math approach: If a tenant represents 8% of revenue, model an 8% revenue hit flowing mostly to NOI (it won't be perfectly linear, but for stress-testing, linear is fine). Then add a one-time cash cost estimate for re-leasing. Even a rough assumption, like 1–2 years of that tenant's rent as a leasing/repositioning cost, can be eye-opening.

If the REIT shrugs off the loss in your model, you've learned something valuable about diversification and cushion. If it breaks, you've learned something even more valuable: the dividend was never the main risk—concentration was.

Stress-testing a portfolio (not just individual REITs)

Investors often run stress tests on single holdings and forget the more humiliating truth: portfolios fail through correlation. In calm times, your REITs may look diversified because they have different names and different property types. In stress, they can all respond to the same macro triggers: rates, credit spreads, and risk appetite.

A useful portfolio stress test asks: *What if my entire REIT sleeve drops 30% in price while dividends are flat or slightly down?* Could you hold? Would you sell? Would you buy more? The correct answer isn't "I'd be fearless." The correct answer is a realistic plan aligned with your temperament.

Another portfolio-level stress is liquidity. If you rely on REIT dividends to fund living expenses, a dividend cut is not just a "lower return." It's an interruption to your cash needs. Stress-testing should therefore include your own household balance sheet: do you have a cash buffer that allows you to avoid selling into a drawdown? Investors love analyzing REIT liquidity and ignore their own.

Turn scenarios into decisions (otherwise it's just doomscrolling with equations)

A stress test should end with action, even if the action is "do nothing, but with conviction." Here are the kinds of decisions stress-testing is meant to provoke:

- If one REIT breaks in every scenario, reduce the position size or replace it.
- If multiple REITs fail the same stress (e.g., refinancing shock), diversify the risk: add REITs with longer maturities, more fixed-rate debt, or stronger coverage.
- If your portfolio is resilient but you personally can't stomach the drawdown, reduce exposure anyway. A portfolio you can't hold is a theoretical portfolio.

Stress-testing is also a great lie detector for "yield logic." If the reason you own a high-yield REIT is income stability, and your stress test shows the dividend is the first thing to crack, the investment thesis was never stable. It was just comforting.

A simple scoring approach: mild, moderate, severe

If you like structure, label scenarios as mild, moderate, and severe. Not because reality comes in neat boxes, but because your thinking should. Mild stress should feel like "annoying but fine." Moderate stress should feel like "management has to work." Severe stress should feel like "only the well-prepared survive."

Then define what "survive" means. For many investors, survival looks like:

- dividend coverage remains adequate or only modestly pressured,
- liquidity remains available,
- no forced equity issuance at distressed prices,

- no covenant-driven panic.

If a REIT fails the severe scenario but passes mild and moderate, that might be acceptable depending on your risk tolerance and position size. If it fails the mild scenario, the market isn't the problem—your underwriting is.

The payoff: calm that is earned, not borrowed

The market will always find new ways to be dramatic. Stress-testing doesn't eliminate drama; it makes it less persuasive. When you've already asked, "What if vacancy jumps?" or "What if interest expense spikes?" the headline loses some of its power. You're not surprised. You're simply watching a scenario you already rehearsed.

That's the quiet advantage of financial wargames: they turn panic into process. They don't make you clairvoyant. They make you prepared. And in a world where capital markets can slam shut with no warning, preparedness is not a personality trait. It's a return driver.

Chapter 5

Constructing a Tax-Efficient Portfolio

You've identified high-quality REITs, but if you buy them in the wrong account, the IRS effectively becomes your silent business partner-taking a much larger cut of your rent checks than necessary. This chapter isn't just about what to buy; it's about where to put it and how to size your bets. We will move from the 'why' of portfolio construction to the specific mechanics of asset location, ensuring your real estate empire benefits you, not the taxman.

5.1 The Job Description: What REITs Do for You

A lot of investing mistakes come from a bad hire.

People buy a REIT the way they buy a new kitchen gadget: because the box promises convenience and the reviews say "prints money". Then the first time interest rates jump, the price falls 20%, headlines scream about commercial real estate, and suddenly the gadget is "broken". It wasn't broken. It was doing the job it was always going to do—just not the job the investor imagined.

So give REITs a job description. Not a slogan. A real job description with duties, strengths, weaknesses, and the kind of work environment where they shine.

A publicly traded REIT is a *real estate operating business* that happens to trade like a stock. It collects rent (or fees that look a lot like rent), pays property-level expenses, pays interest on its debt, reinvests to keep buildings competitive, and then distributes a large share of what's left to shareholders. That last part—the habit of paying out cash—is why REITs keep showing up in income conversations. But the rest of the story matters just as much: REITs are not "real estate" in the abstract. They're businesses with customers, leases, competition, labor costs, refinancing risk, and the occasional tenant that goes from "blue-chip" to "please don't bounce that check".

When you hire REITs for what they're actually good at, they can be one of the most psychologically helpful holdings in a portfolio. Not because they eliminate volatility—they don't—but because they can make your return stream feel less imaginary. A REIT portfolio can turn the stock market from a scoreboard into a set of cash-producing assets. You're less tempted to measure success only by today's price quote when you can point to occupancy, lease escalators, and rent checks arriving quarter after quarter.

That doesn't mean REITs are a substitute for growth stocks. It means they can be a different engine in the same machine. Growth stocks are often hired to compound capital by reinvesting profits internally. Many REITs are hired to *distribute* a meaningful portion of the profits and let you decide what to do with the cash—spend it, rebalance, buy something else, or reinvest into the REIT itself. Those are different jobs, and confusing them is like hiring a firefighter and getting mad that they don't cater weddings.

The first duty on the REIT job description is *cash flow you can actually use*. REITs tend to pay higher dividends than the broad stock market because the structure pushes them toward distribution. That's the headline. The more useful detail is that REIT income often behaves differently from the dividend stream of, say, mega-cap consumer brands. A classic "dividend stock" might raise its payout slowly and predictably while most of your return comes from price appreciation. A REIT's total return often feels more like a two-part paycheck: a visible cash component plus a price component that can be moody, sometimes dramatic, and frequently tied to interest rates and credit conditions.

This is where expectations need to be specific. The cash is not a magical coupon clipped from thin air. It's your share of a business that owns depreciating buildings and constantly fights entropy. Roofs wear out. Parking lots crack. HVAC systems die at the worst possible time. A good REIT replaces and upgrades these things without turning the distribution into a roller coaster. A great REIT does that while also finding smart ways to grow—acquisitions at good prices, development where it has an edge, and leases that steadily mark rents to inflation. But even the great ones have years when the market doesn't feel like paying up for that competence.

The second duty is *inflation-aware income*. "Inflation hedge" gets tossed around like a universal truth, so let's make it concrete. Many leases have built-in rent escalators—some fixed (2% per year), some tied to CPI, some tied to market resets. Apartments can reprice quickly because leases are short. Hotels can reprice nightly. Self-storage can reprice fast in a hot market. Industrial leases might be longer but can include bumps and can reset meaningfully at renewal. The point is not that REITs automatically win during inflation. The point is that many property types have *mechanisms* that can push revenue upward when prices in the economy rise.

But inflation has a dark twin: higher interest rates. If inflation rises and the central bank responds by pushing rates up, REITs can get hit in two ways. First, higher rates can raise borrowing costs when debt matures and must be refinanced. Second, higher rates often raise the required return on income-producing assets, which can compress property values and, by extension, REIT share prices. That's why you'll sometimes see the market punish REITs even when the buildings are full and the rent checks are arriving on schedule. The job includes inflation-awareness, not immunity from the interest-rate cycle.

The third duty is *exposure to a different set of economic levers*. Broad equity markets are dominated by businesses whose raw material is code, branding, or intellectual property. REITs are dominated by physical constraints: location, zoning, replacement cost, and the slow, stubborn reality that it takes time and permits to build stuff. That difference matters because it means REITs can zig when other equities zag. Not always, and not on your preferred timeline, but enough to be useful.

A simple way to feel this is to think about what makes a great REIT year. It's rarely a viral product launch. It's usually a combination of boring, powerful forces: rent growth, improving occupancy, disciplined expense control, and a cost of capital that allows accretive acquisitions. Those are not the same drivers that power a software company. When you blend different drivers, you reduce the chance that one macro shock hits everything in your portfolio with equal force.

Of course, REITs are still stocks. That matters more than many investors expect. When panic hits markets, correlations often rise and everything sells off together—especially anything perceived as "risky" or "rate-sensitive". If your mental model is that REITs are a calm, rental-like holding that will never drop 30%, you've hired them for the wrong job and you'll fire them at exactly the wrong time.

So here's a better mental model: REITs can be *income-focused equities with real-asset DNA*. They can deliver long-term wealth building through a mix of income and growth, but the ride can be as bumpy as any other equity category. The cash flow is the stabilizer, not the price chart.

It also helps to separate two kinds of volatility: *business volatility* and *market volatility*. Business volatility shows up in the numbers—occupancy, rent collections, same-store net operating income, funds from operations, dividend coverage. Market volatility is the mood swing of the trading screen. REIT investors get whiplash when they treat market volatility as a direct report card on the business. In reality, a REIT can have a great quarter operationally and still see the stock drop because the 10-year Treasury yield moved, credit spreads widened, or a different REIT blew up and tainted the whole sector by association.

One of the healthiest things a REIT can do for you is force you to pay attention to *how* returns are generated. With a broad index fund, you can mostly ignore the plumbing and still do well. With REITs, the plumbing is the point. You start noticing how leverage works, why debt maturity schedules matter, and how a tenant roster can be a hidden risk factor. You begin to think like an owner, because you actually *are* one.

But to keep this job description honest, we have to name the parts of the work REITs are not good at.

REITs are not hired for explosive, uninterrupted growth. Some REITs do have powerful growth runways—data centers, towers, industrial in certain cycles—but the structure naturally pushes cash out the door. A company that pays out a large share of its cash flow has less internally retained capital for moonshot reinvestment. That can be a feature for investors who want cash discipline. It can also be a limitation if you expect the return profile of a company that reinvests aggressively and pays little or no dividend.

REITs are also not hired to be "safe" in the way people mean it at the dinner table. The buildings can be high-quality, the tenants can be strong, and the leases can be long—but the stock can still be volatile. If you need something that won't meaningfully decline in a bad year, that's not a REIT's job. That's the job of cash, short-term Treasuries, or high-quality bonds (with their own caveats). A REIT is an equity claim on a leveraged asset base, and equity claims wiggle.

And REITs are not hired to be effortless. You can absolutely buy a diversified REIT fund and keep life simple, but the category itself demands respect. Property type matters. Lease structure matters. Balance sheet design matters. A REIT with floating-rate debt and near-term maturities is living in a different universe than a REIT with long-dated fixed-rate debt. A REIT that must constantly access capital markets to grow has a different stress profile than one that can fund growth internally. If you don't want to learn any of this, that's fine—just be honest that you're outsourcing the job to a fund manager or index provider.

Now, let's talk about the emotional job REITs do, because portfolios are built in the real world, where humans have nervous systems.

A lot of investors say they want maximum return, but what they actually need is *maximum return they can stick with*. REITs can help here in a weird way: the dividend creates a rhythm. Even when the price is ugly, the distribution can feel like proof that the underlying machine is still running. That can reduce the temptation to sell at the bottom—provided you're not using the dividend as a blindfold.

The dividend can also act like an automatic rebalancing tool. If you reinvest dividends, you're buying more shares when prices are down and fewer when prices are up. That's not magic; it's just math and discipline packaged as a default setting. If you take the cash instead, it can fund other purchases or reduce the need to sell other assets for living expenses. Either way, the job is *flexibility*—a stream of cash that gives you options.

But there's a trap hiding inside that same paycheck: yield-chasing. REIT yields can spike when prices fall. Sometimes the market is wrong and you're being paid to be brave. Other times the market is correctly sniffing out a dividend cut, a refinancing problem, or a tenant base that's about to crater. A REIT's dividend yield is not a savings account interest rate. It's a market signal. Treat it like one.

If you want a clean one-sentence job description, use this:

A REIT is hired to convert real-world property economics into a reliable (but not guaranteed) stream of cash flow, plus a long-term shot at appreciation—while accepting equity-like volatility, especially around interest-rate cycles.

That sentence bakes in the key trade: you're getting income and real-asset exposure, but you're paying for it with price turbulence and the need for patience.

It also sets up a smarter way to evaluate success. For many REIT investors, success is not "my REIT is always up". Success is "my REIT sleeve kept paying me through the cycle, and the long-term total return justified the stomach knots". You judge it the way you'd judge a rental property you own indirectly: did the income hold up, did the asset stay competitive, did the financing stay sane, and did you avoid catastrophic mistakes?

Another part of the job description is *portfolio counterweight,* but with a footnote. REITs don't always diversify in the way people hope. In some selloffs, they drop alongside the market. In rate spikes, they can drop harder. Diversification isn't a promise that one thing goes up when another goes down; it's a reduction in the chance that everything breaks for the same reason at the same time. REITs can still help because their underlying cash flows are tied to rent and occupancy rather than ad revenue, iPhone cycles, or enterprise software budgets. But the stock market can temporarily ignore that nuance and trade REITs like a duration-heavy bond proxy.

The practical takeaway is to expect REITs to behave like *their own messy category*. Sometimes they look like value stocks. Sometimes they look like long-duration assets. Sometimes they look like a niche credit story. And sometimes they just look like whatever narrative is loudest on financial television that week.

If that sounds unsettling, it's actually liberating. You stop demanding that REITs act like a specific stereotype. You start letting them be what they are: publicly traded claims on a collection of property businesses. The less you force them into the wrong role, the more useful they become.

There's also a subtle but important difference between buying REITs for *income* and buying them for *spending*. Income is the cash the investment produces. Spending is your decision about what to do with it. A retiree might hire REITs specifically to cover part of monthly expenses. A younger investor might hire REITs to generate cash that gets reinvested into other parts of the portfolio, essentially using REIT dividends as a rebalancing fuel. Same holding, different job assignment.

That job assignment should be written down somewhere—not as a formal investment policy statement unless you enjoy that sort of thing, but as a personal rule you can read when the market is screaming. Something like: *"I own REITs because I want a steady stream of cash flow and exposure to property economics. I expect the price to be volatile. I will judge this holding on multi-year income stability and long-term total return, not on whether it outperforms the S&P 500 every quarter."*

If you can't say something like that with a straight face, you're not ready to hire REITs yet. Or you're ready, but only through a broad fund so the category can do its work without you micromanaging it.

One more expectation to set: REITs can be *boring in a productive way*. The best-run REITs often look unremarkable on the surface. They own assets in places with constrained supply. They keep leverage at levels that won't force a fire sale. They sign leases that steadily nudge rents upward. They raise the dividend gradually. They avoid dramatic stories. In a market that rewards drama, that can feel underwhelming—until a recession hits and the boring REIT is still standing while the exciting one is doing emergency equity issuance at a price that makes shareholders wince.

That's not a promise that "boring always wins". It's a reminder that the job you want done—steady cash flow with survivability—usually comes from habits that don't make good headlines.

Finally, keep the job description rooted in total return reality. REIT investing is not only about dividends. The dividend is the *delivery mechanism* that makes the return tangible, but the long-term outcome still depends on what you paid, how the cash flow grows, and what valuation the market assigns to that cash flow over time. A high dividend can be a gift. It can also be a distraction from weak fundamentals. A low dividend can be a sign of a growth-oriented REIT reinvesting more cash. It can also be a sign of trouble. The job is not "high yield". The job is "convert property economics into shareholder returns in a way that fits my portfolio."

When REITs are hired with that clarity, they tend to do something quietly powerful: they make your portfolio feel more like a set of productive assets and less like a collection of lottery tickets. You're not betting on a future narrative. You're owning cash-flowing businesses tied to places people live, store goods, send data, park cars, see doctors, and buy groceries. The market can be dramatic about that reality, but the reality itself is stubborn.

And stubborn, in investing, is often exactly what you want on the team.

5.2 The Goldilocks Zone: Sizing and Diversification

REITs have a peculiar talent: they can make you feel diversified while quietly concentrating your risk.

Own five different tickers and you might still be making the same bet five times—on the same kind of tenant, the same kind of financing, the same kind of economic weather. That's how investors end up *surprised* by losses that were, in hindsight, extremely predictable. Not because REITs are inherently dangerous, but because real estate is lumpy. One property type can be a gentle cash machine for a decade and then get kneecapped by a single structural shift: remote work, new supply, a credit crunch, a technological leap, a demographic swing.

Sizing and diversification are how you turn REITs from a thrilling side quest into a durable part of a portfolio. The goal isn't to avoid drawdowns—equities draw down. The goal is to avoid the kind of drawdown that changes your behavior: panic selling, vow-making, and the sudden conviction that "this whole REIT thing was a scam." Catastrophic losses usually aren't caused by one bad quarter. They're caused by the same mistake repeated at full volume: too much in one name, too much in one sector, too much exposure to one macro variable, and too little humility about what you *don't* know.

Start with a simple question: how much real estate do you want in your life?

The best allocation number is the one you can hold through ugly headlines. For many generalist investors—people whose portfolios also include broad stock funds and bonds—REITs often land in a *meaningful-but-not-dominant* range. In plain English: enough to matter, not enough to sink you.

As a rule of thumb, think in the neighborhood of 5%–15% of the total portfolio for a "REIT sleeve." That range is popular for a reason. At 5%, you'll feel the income and diversification benefit without turning your financial life into a referendum on commercial real estate. At 15%, REITs can materially boost portfolio yield and real-asset exposure, but the portfolio's mood will start to track interest-rate cycles more closely. Above that, you're no longer "adding REITs." You're making a statement about real estate as a dominant asset class in your financial plan.

The right number depends on what you're hiring REITs to do. If they're meant to be a supporting actor—income, inflation-aware cash flows, some diversification—then a supporting-actor allocation makes sense. If you're building a portfolio around income from day one, you might justify the high end of that range. If you already own significant real estate outside the stock market—your home, a rental, a family property partnership—your *true* real estate exposure is probably higher than your brokerage account suggests, and the stock REIT allocation should be sized with that in mind.

One more reality check: a REIT allocation isn't just a pie-slice decision. It's a *tax decision* wearing an allocation costume. Put a high-yield REIT allocation in a taxable account and you may discover that the portfolio produces lots of income—most of which is immediately shared with the IRS. Put the same allocation in a tax-sheltered account and the compounding can feel meaningfully different. The "how much" question is inseparable from the "where" question, even if they get discussed separately.

The concentration trap: when "diversified" is a costume

In real estate, sector concentration is the most common way investors accidentally step on a landmine.

Office is the obvious example because the post-2020 pain has been loud and public. But concentration traps exist everywhere. A portfolio heavy in malls can look safe right up until tenant bankruptcies accelerate. A portfolio heavy in healthcare can look defensive right up until reimbursement changes or labor costs bite. A portfolio heavy in self-storage can look unstoppable right up until supply catches up and rent growth stalls. A portfolio heavy in industrial can look like a permanent winner right up until a manufacturing slowdown or overbuilding changes the balance of power.

The reason concentration is so seductive is that it usually begins as a success story. One sector performs well, dividends look steady, the narrative is comforting, and investors gradually "average up" into a single theme. Then the environment changes, and what felt like conviction starts to look like a lack of imagination.

The trick is to diversify across property economics, not just ticker symbols. "Different" REITs can share the same heartbeat:

- The same tenant base (for example, the same handful of big-box retailers or the same class of venture-backed tech firms).

- The same lease structure (short leases that reprice quickly are great—until the repricing works against you).
- The same financing dependence (capital-hungry REITs can suffer when markets shut).
- The same sensitivity to interest rates (long-duration cash flows get repriced when yields move).

If you diversify across those underlying drivers, you lower the chance that one shock hits your entire REIT sleeve like a hammer.

A picture worth a thousand ``I swear I'm diversified" conversations

Place a visual near the point where investors usually fool themselves: when they own multiple holdings but one sector quietly dominates. The following chart makes the problem obvious at a glance.

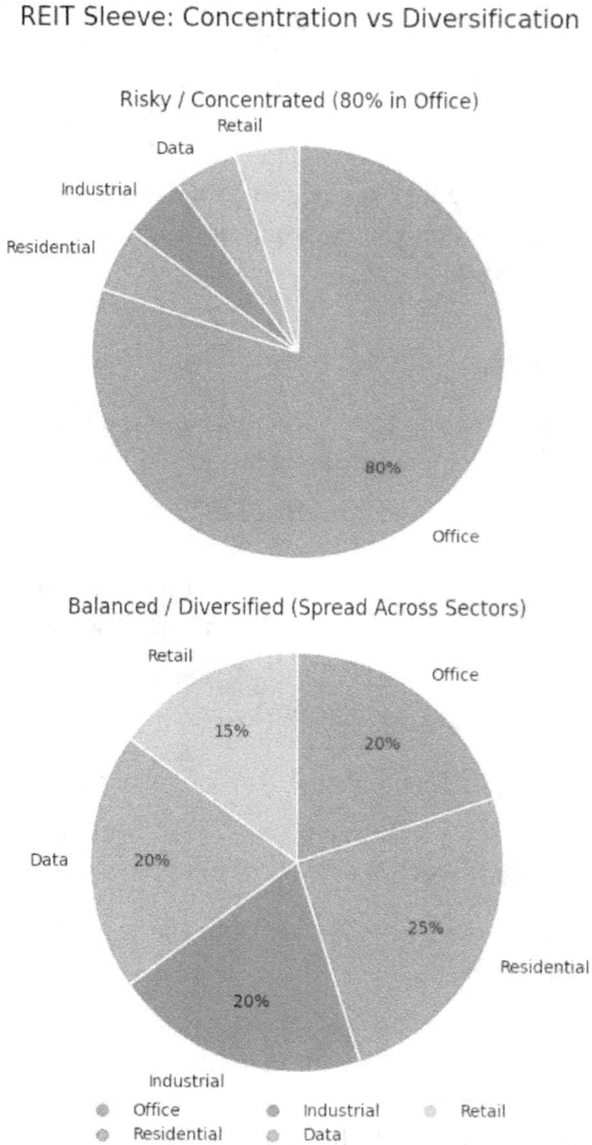

If that top pie makes you flinch, good. It should. An 80% bet on one property type is not "a REIT allocation." It's a thesis with a portfolio attached. Sometimes a thesis is right. The problem is that a thesis can be wrong for years without technically being *dead*. Office buildings can keep collecting rent while their long-term value erodes. Malls can pay dividends while tenant quality deteriorates. A concentrated portfolio can lull you into complacency until the refinancing wall arrives, or a big tenant leaves, or a wave of new supply turns yesterday's pricing power into today's concession-heavy scramble.

Position sizing: the ``one bad tenant" rule

The fastest way to turn a REIT allocation into a stomach ache is to let any single name become a make-or-break holding.

A practical position-sizing rule: if one REIT implodes—dividend cut, dilutive equity raise, major tenant bankruptcy, debt downgrade—you want the damage to be annoying, not life-changing. That usually means keeping individual REIT positions to a low single-digit percentage of the total portfolio.

Many investors find sanity in limits like:

- No single REIT more than 2%–4% of the total portfolio.
- No single REIT sector more than 25%–35% of the REIT sleeve (which itself is only a slice of the full portfolio).

The exact numbers are less important than the principle: build in the ability to be wrong.

Because you *will* be wrong. Not because you're careless, but because real estate has more moving parts than stock charts suggest. Tenants merge. Cities change. Interest rates do their thing. A CEO makes an acquisition that looks brilliant on a slide deck and painful on a balance sheet. A "safe" REIT discovers that its biggest customer was quietly shopping for cheaper space.

A position limit is a behavioral guardrail. It stops your portfolio from becoming a referendum on whether you correctly predicted the future of, say, suburban office parks or skilled nursing reimbursements.

Diversification that actually matters: drivers, not labels

"Diversify by sector" is a good start, but it's not the finish line. Two REITs in different sectors can still be tightly linked if they share the same vulnerability.

A more durable approach is to diversify across *return drivers*:

- *Lease duration*: Short leases (apartments, hotels, self-storage) can reprice faster; long leases (net lease, some industrial) can provide stability but may lag inflation.
- *Tenant health*: Consumer-dependent tenants behave differently from enterprise tech tenants, and both behave differently from government or healthcare payors.
- *Balance sheet structure*: Lots of near-term maturities versus laddered, long-term fixed-rate debt is the difference between "annoyed by rates" and "defined by rates."
- *Supply dynamics*: Some assets are easy to build (a new self-storage facility); others are hard (entitled urban infill, certain specialized facilities). Easy-to-build sectors can go from boom to glut fast.

If you spread exposure across those axes, you'll still have drawdowns, but you're less likely to wake up to a portfolio where *every* holding is being punished for the same structural reason.

This is also where broad REIT funds earn their keep. A diversified fund can give you exposure across many property types without forcing you to become an expert in each one. The tradeoff is that you accept the sector weights the market gives you, including exposure to sectors you might not personally love. That's not a flaw; it's the price of simplicity.

The hidden concentration: interest-rate sensitivity

Even diversified REIT sleeves can share one big, invisible risk factor: they can all react to the same move in rates.

This happens for two reasons. First, REIT cash flows are often valued like long-lived income streams. When the "risk-free" rate rises, the market often demands a higher return from those income streams, which can mean lower prices today. Second, REITs use debt. Higher rates can raise future interest expense, which can pinch cash available for dividends and growth.

The practical implication is not "avoid REITs when rates are rising." Market timing is a seductive hobby with terrible long-term results. The practical implication is to diversify *within* REITs by balance sheet resilience. When you look at a REIT, you want to know: does it *have to* refinance a big chunk of debt soon, or does it get to choose its moments? That question often matters more than whether the properties are nice.

A REIT with strong properties and weak financing can still hurt you. A REIT with merely good properties and excellent financing can survive ugly cycles and come out with opportunities. Real estate is not just about buildings; it's about *capital access*.

Sizing the REIT sleeve when you already own real estate

This is where people get unintentionally funny: they'll say they want "some real estate exposure" while sitting in a house that represents 40%–70% of their net worth.

Your primary home is not a REIT. It doesn't send you a dividend and it has a different risk profile (local market risk, maintenance, insurance surprises, property tax policy, personal liquidity constraints). But it *is* real estate exposure—often heavily leveraged exposure. If you also own a rental property, you've added tenant risk, vacancy risk, and more leverage sensitivity.

So when you add a public REIT allocation on top, you're layering real estate on real estate. That can still be a great decision, but it should be a *conscious* one. Many investors should think of REITs not as "my only real estate" but as "my liquid, diversified real estate businesses" sitting beside a very concentrated, illiquid real estate position (their home).

This framing can keep you from doing something common and painful: over-allocating to REITs because the dividend feels comforting, while ignoring that your biggest single asset is already a giant bet on one zip code.

A practical blueprint: build a REIT sleeve that can take a punch

A punch-resistant REIT sleeve doesn't require perfection. It requires a few constraints that prevent catastrophic outcomes.

One approach that works for many investors:

- Keep the total REIT sleeve in a range where you can sleep (often 5%–15% for generalists).
- Prefer broad diversification across property types, either via a fund or a small basket of REITs deliberately chosen to be different.
- Cap single-name exposure so one dividend cut can't derail your plan (low single-digit percent of total portfolio).
- Watch sector concentration like a hawk, especially after a sector has had a great run (winners become overweight without asking permission).

Notice what's missing: a promise that you'll avoid losses. Losses happen. The point is to avoid *ruin*. Ruin is what occurs when you make a handful of unforced errors at the same time: too big, too concentrated, too yield-hungry, too confident that "this sector is different."

The most underrated benefit of good sizing: it lets you hold through the ugly parts

REITs tend to get tested at the worst possible moments—right when you feel most foolish for owning them. That's usually when the long-term investor has the biggest advantage, because the market often prices REITs with a short-term emotional discount. But you only get paid for that patience if your sizing allows you to stay put.

If your REIT sleeve is so large that a drawdown forces you to sell, it's not an "allocation." It's a fragility. If your single-name position is so big that a dividend cut feels like a personal betrayal, it's not "conviction." It's overexposure.

The Goldilocks zone—*not too big, not too concentrated, not too clever*—is what turns REIT investing into a repeatable process. You can be wrong about a tenant, wrong about a sector, wrong about the timing of rate cuts, and still come out fine. That's the whole point of portfolio construction: not to predict the future, but to remain investable no matter what the future does.

5.3 The Tax Trap: Location, Location, Location

A REIT can be a wonderful cash machine and a terrible tax machine at the same time. The frustrating part is that both statements can be true *even when the REIT performs exactly as hoped*. The buildings are occupied, rents rise, the dividend arrives on schedule—and a big chunk of that cash quietly leaks out through taxes before you can reinvest it, rebalance with it, or spend it.

That leak is called *tax drag*: the slow, compounding damage caused by paying taxes along the way instead of letting returns grow untouched. Tax drag is especially relevant for REITs because they are built to distribute a lot of cash. A high-distribution asset placed in the wrong account can feel like trying to fill a bathtub with the drain open.

Why REIT dividends get ``ordinary" treatment so often

Investors come into REITs carrying a perfectly reasonable assumption: "Dividends are dividends." Then they meet the U.S. tax code's personality quirks.

Many stock dividends are *qualified dividends*, meaning they may be taxed at the lower long-term capital gains rates (assuming you meet certain holding-period and other requirements). REIT dividends, by contrast, are frequently *not* qualified. In plain language, that means much of the dividend often gets taxed more like paycheck income than like a preferentially taxed stock dividend.

That difference matters because the spread between ordinary-income rates and qualified-dividend rates can be wide. When you're compounding over years, a few percentage points of annual tax drag is not a rounding error—it's a different ending.

It's not that REITs are "punished." It's that the REIT structure is designed to pass income through to shareholders. You're receiving income that looks a lot like rental profit, and the tax system often treats it accordingly.

The sneaky twist: a REIT dividend isn't always one thing

Even when REIT dividends are mostly ordinary income, they're not always *entirely* ordinary income. REIT distributions can come in different flavors, and your broker reports these flavors on Form 1099-DIV.

Common categories you'll see:

- *Ordinary dividends*: usually the big one for REITs, typically taxed at ordinary income rates.
- *Capital gain distributions*: taxed like long-term capital gains.

- *Return of capital (nondividend distributions)*: not taxed immediately, but it reduces your cost basis (which can increase your taxable gain when you sell).

That third category is where investors get whiplash. Return of capital can feel like a "tax-free dividend," but it's better described as *tax-deferred*. The IRS isn't necessarily waving goodbye; it may simply be waiting near the exit.

Still, the headline remains: REITs tend to throw off a lot of current taxable income compared with many other equity investments. That's the heart of the asset-location problem.

Asset location: the most boring move that can boost real results

Asset allocation is what you own (stocks, bonds, REITs). *Asset location* is where you own it (taxable brokerage, Traditional IRA/401(k), Roth IRA/401(k), HSA, etc.). People obsess over allocation because it's exciting and visible. Location is less glamorous—until you run the numbers and realize it can meaningfully change what you keep.

A useful rule of thumb is simple: put tax-inefficient, income-heavy assets in tax-sheltered "containers" when you can, and put more tax-efficient assets in taxable accounts. Many REITs fall into the "tax-inefficient" bucket for one reason: they often produce taxable income year after year.

That doesn't mean REITs *must* be held in an IRA. It means you should assume the taxable account is guilty until proven innocent.

Choosing the container: Traditional vs Roth vs taxable

Think of accounts as different kinds of deal-making with the IRS:

Traditional IRA/401(k) is a deferral deal. You generally avoid taxes today, the money compounds without annual tax bills, and you pay taxes when you withdraw.

Roth IRA/401(k) is a prepayment deal. You pay taxes up front (on contributions), then qualified withdrawals can be tax-free, and the compounding happens in a near-perfect greenhouse.

Taxable brokerage is the open-air market. You can come and go easily, harvest losses, donate appreciated shares, and access money without retirement rules—but you pay taxes along the way.

For REITs, the big temptation is to focus on the dividend yield and forget the container. A 6% yield in a taxable account is not the same as a 6% yield in a Roth. In taxable, every year is a toll booth. In a Roth, the compounding can be dramatically cleaner.

A Traditional account can also be a strong home for REITs because it shelters the ongoing distributions from annual taxation. The tradeoff is that withdrawals are generally taxed as ordinary income, and required minimum distributions (RMDs) can force money out later. A Roth avoids RMDs in many cases and can be especially appealing for high-growth assets, but many investors reasonably like placing high-current-income assets (like many REITs) into Traditional accounts to reduce annual tax drag.

No single placement is universally "correct." The point is to make the choice on purpose rather than by accident.

The Section 199A deduction: the REIT tax discount that people half-remember

REITs come with a partial consolation prize: many investors can take a deduction on *qualified REIT dividends* under Section 199A (often described as a 20% deduction, subject to rules and limitations). Your 1099-DIV frequently reports this in a dedicated place (commonly Box 5).

This is one of those tax features that is both valuable and misunderstood.

It does *not* magically turn REIT dividends into qualified dividends. It does *not* eliminate the tax drag. What it can do—when you're eligible—is lower the effective tax rate on that portion of the REIT income in a taxable account. In other words, it can make the taxable-account penalty smaller, not make it disappear.

Also, eligibility can depend on holding-period rules and your broader tax situation. The tax code loves fine print. If you're relying on 199A to justify large REIT exposure in a taxable account, treat that reliance like you would treat a variable-rate loan: useful, but not a foundation you build your whole house on.

NIIT: the extra 3.8% that high earners forget to include

For higher-income households, the *Net Investment Income Tax* (NIIT) can add another 3.8% layer on certain investment income once modified adjusted gross income exceeds thresholds. That can apply to REIT dividends in a taxable account, and it can apply to capital gains too.

NIIT is a perfect example of why investors should be careful with casual statements like "my bracket is 24%." For some households, the relevant marginal rate on investment income is more like "my bracket plus 3.8%," at least on the portion that falls into NIIT territory. When the income stream is large and persistent—as REIT income often is—that extra layer can meaningfully change the after-tax outcome.

When taxable might still make sense

Sheltering REITs in retirement accounts is a strong default, but life is messy and portfolios are constrained by what you actually have available.

Taxable can still be rational when:

- Your tax-advantaged space is limited and you need it for something even less tax-efficient (for some investors, taxable bonds compete for that space).

- You value liquidity and flexibility more than you dislike annual tax bills.
- You're in a low bracket now and expect higher taxes later, making taxable income less painful today than future ordinary-income withdrawals.
- The REIT's distribution includes significant return of capital at times (still not "free," but it can reduce current-year tax).

Even then, it's worth treating taxable REIT ownership as a decision that should earn its keep. If a REIT is going to live in taxable, you want a reason beyond "I ran out of room" or "the yield looked nice."

The practical decision tree (the one you wish every brokerage app popped up)

Right in the middle of any serious REIT plan, it helps to have a quick-reference tool that forces the right questions. The flowchart below is intentionally simple. It doesn't pretend to capture every exception in tax law. It's trying to prevent the most common mistake: placing a high-current-income asset into a taxable account by default, then acting shocked when April arrives.

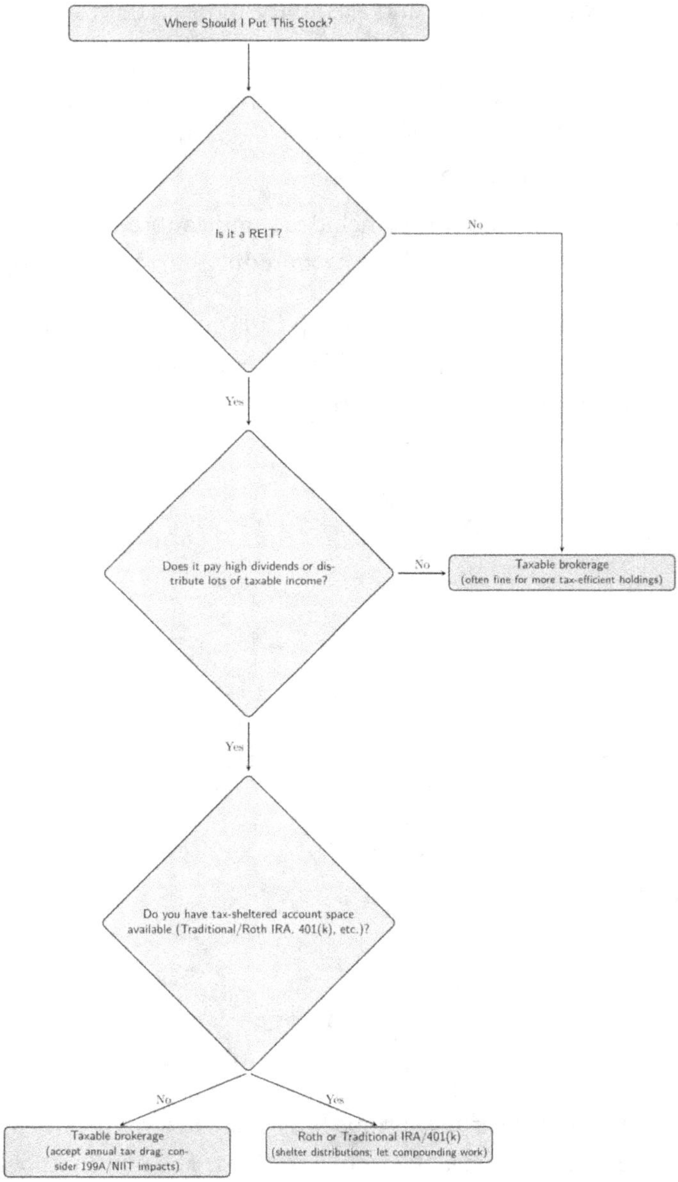

The flowchart's bias is deliberate: if something reliably spits out taxable income, you want a good reason to leave it exposed.

A concrete example: the ``same REIT, different life'' problem

Two investors buy the same REIT at the same price and the same dividend yield. One holds it in a taxable account. One holds it in a Roth IRA.

Both investors receive the same cash distribution from the REIT. Only one gets to keep all of it working for them. The taxable investor loses a portion each year to taxes (partly offset in some cases by 199A, sometimes increased by NIIT). The Roth investor reinvests every dollar without annual tax friction, assuming qualified Roth rules are met.

Over one year, the difference can look modest. Over a decade, the difference can become a second shadow portfolio—one built out of all the dollars that would have gone to taxes but instead stayed invested.

That's the quiet cruelty of tax drag: it doesn't announce itself with a crash. It simply lowers your ceiling.

The psychological trap: ``But I like getting paid''

REIT investors often have a healthy preference for cash flow. The trap is letting the *feeling* of getting paid override the math of keeping it.

A high dividend in taxable can create a comforting routine—until you realize you're effectively doing forced, taxable withdrawals every year whether you need the cash or not. If you're reinvesting, you're reinvesting *after* taxes, which means you're compounding a smaller base than you could have.

There's a time and place for taxable dividends (especially when you truly want to spend the income), but if the plan is "reinvest for the long run," then paying avoidable annual tax bills is like paying rent on your own money.

A few placement heuristics that actually hold up in real life

The tax code is complicated, but the portfolio decisions don't have to be.

Reasonable heuristics many investors can use:

- *If the asset reliably produces ordinary-looking income, try to shelter it.* Many REIT distributions fit here.
- *If the asset's return is mostly price appreciation and/or qualified dividends, taxable is often more forgiving.*
- *If you're forced to hold REITs in taxable, be extra intentional about sizing.* Tax drag scales with the size of the position.

These aren't iron laws. They're guardrails. They reduce the odds that you build a portfolio that looks great on a gross-yield basis and quietly disappoints on a net-of-tax basis.

One last reality check: tax rules change, but the strategy is durable

Tax rates and deductions can change with legislation. Details around deductions like 199A can evolve. Thresholds can move. The exact boxes on a form can get renumbered. That's all real.

But the core insight is sturdier than any one tax year: *when an investment sends you taxable income every year, the account you choose becomes part of the investment's return.* For REITs, "location" is not a cute pun. It's a performance variable.

Treat the container as part of the asset. Put the cash machine where it won't be forced to share its output with the tax collector every single year. That's not a loophole. It's simply using the rules as they were designed: taxable accounts for flexibility, tax-sheltered accounts for compounding, and REITs placed with enough intention that their biggest risk isn't the IRS.

5.4 The Bottom Line: Calculating After-Tax Returns

A dividend yield is a *sticker price*. It's what the seller wants you to look at while you're falling in love with the purchase.

What you actually live on—what you can reinvest, spend, or rebalance with—is the *after-tax* yield. That's the number that shows up in your checking account, or the number that remains inside your IRA after the dust settles. Two investments can advertise the same yield and deliver wildly different real-world outcomes once taxes take their cut.

This is where REIT investing gets wonderfully concrete. You don't have to guess at the tax character of your distributions. The brokerage doesn't send you a poem about it; it sends you a form. If you learn to read that form like an owner instead of like a bystander, you stop chasing shiny yields and start comparing investments the way a serious allocator would: by *net* results.

Headline yield vs. take-home yield: the difference between marketing and math

If a REIT yields 6%, it is not paying you 6%. It is paying you 6% *before* your tax situation gets involved.

In a tax-sheltered account, the difference between headline and take-home can be close to zero in the current year (though Traditional accounts generally tax you later upon withdrawal). In a taxable brokerage account, the gap can be immediate and persistent—especially for investors in higher brackets, and especially when the distribution is largely taxed as ordinary income.

That gap is tax drag's favorite disguise. It rarely feels dramatic in one year. It feels like "a little annoyance." Over ten years, it can be the difference between a portfolio that compounds like a machine and one that compounds like a machine with sand in the gears.

The four buckets that matter (and the one investors forget)

REIT distributions show up to you as cash, but the IRS cares about *what kind* of cash. In taxable accounts, the same dollar amount can be taxed in very different ways depending on its character. Your 1099-DIV is essentially a sorting hat.

For after-tax return calculations, the most useful way to think about REIT payouts is as four buckets:

- *Ordinary income portion:* usually the main component. Taxed at ordinary income rates (and potentially subject to NIIT for higher earners).
- *Qualified dividends (sometimes):* less common for REIT common dividends, but can appear in certain circumstances.
- *Capital gain distributions:* taxed at long-term capital gains rates.
- *Return of capital (ROC):* not taxed right away, but reduces your cost basis, which can increase the taxable gain when you sell.

That last one—return of capital—is the bucket investors forget, misunderstand, or treat like a free lunch.

Return of capital can be genuinely beneficial *in the present* because it defers taxes. But it changes the future by lowering your basis. Lower basis means more gain later if you sell. ROC is less "tax-free dividend" and more "tax postponed until you cash out." Still useful, just not magical.

A realistic shortcut: using expected tax character instead of wishful thinking

Every REIT's distribution mix varies by company and by year. Some years have more capital gains; some years have more return of capital; some years are heavy ordinary income. But if you want a realistic illustration without pretending you can predict your next 1099-DIV perfectly, it's reasonable to use a rule-of-thumb split that resembles what many investors have seen in aggregate in recent years: mostly ordinary income, with smaller slices of capital gains and return of capital.

The value of this shortcut isn't precision. It's honesty. It prevents the most common self-deception: treating a REIT distribution as if it's taxed like a qualified dividend stock just because both show up under the word "dividend."

Tax-equivalent yield: the only yield comparison that isn't lying to you

When investors compare yields across investments with different tax treatment, they're often comparing apples to oranges and declaring victory.

A cleaner comparison tool is *tax-equivalent yield*: the pre-tax yield an investment would need to offer to match the after-tax yield of another investment.

For a simple case where an investment's distribution is taxed at a single effective rate, the math is straightforward:

$$\textit{After-tax yield} = \textit{Pre-tax yield} \times (1 - \textit{tax rate})$$

and therefore

$$\textit{Pre-tax yield needed} = \frac{\textit{After-tax yield target}}{(1 - \textit{tax rate})}.$$

The moment you learn to do this, you start seeing "high yield" differently. A high yield in a taxable account can be nothing more than a high *gross* number that gets shaved down every year. A modest yield inside a Roth can quietly win because it keeps all its fuel.

A worked example (without pretending your tax return is identical to anyone else's)

Assume you're considering a REIT yielding 6% in a taxable brokerage account. For illustration, suppose the distribution breaks down like this:

- 80% ordinary income
- 10% long-term capital gain distributions
- 10% return of capital

Now assume a simplified tax situation:

- Ordinary income rate: 32%
- Long-term capital gains rate: 15%
- NIIT: not applied in this example (we'll handle it separately)

Step 1: Convert the 6% yield into its components:

$$\begin{aligned} \textit{Ordinary portion} &= 6\% \times 0.80 = 4.80\% \\ \textit{LTCG portion} &= 6\% \times 0.10 = 0.60\% \\ \textit{ROC portion} &= 6\% \times 0.10 = 0.60\% \end{aligned}$$

Step 2: Apply taxes to the taxable portions:

$$\begin{aligned} \textit{After-tax ordinary} &= 4.80\% \times (1 - 0.32) = 3.264\% \\ \textit{After-tax LTCG} &= 0.60\% \times (1 - 0.15) = 0.510\% \end{aligned}$$

Return of capital is not taxed today in this simplified view:

$$\textit{After-tax ROC (today)} = 0.60\%$$

Step 3: Add them up:

$$\textit{Take-home yield (current year)} \approx 3.264\% + 0.510\% + 0.60\% = 4.374\%.$$

So the "6% yield" behaves more like a 4.37% take-home yield in the current year under these assumptions. That's not a small haircut. It's the difference between compounding quickly and compounding with a permanent headwind.

And it gets more interesting: the 0.60% ROC is not a gift; it's a time-shift. You'll likely pay for it later via higher capital gains when you sell (unless you never sell, donate the shares, or pass them on with a step-up in basis, depending on the rules in effect and your personal situation).

The point isn't to fear taxes. The point is to stop pretending they're not part of the return.

Now add NIIT (because some investors effectively pay a surcharge)

For investors above NIIT thresholds, investment income in taxable accounts may be subject to an additional 3.8% tax. In many practical situations, it's helpful to think of NIIT as increasing the effective tax rate on the ordinary-income portion of REIT distributions (and sometimes on capital gains as well).

Using the same example, if NIIT applies to the ordinary portion, the effective rate could look like 32% + 3.8% = 35.8% on that portion. Then:

$$\textit{After-tax ordinary (with NIIT)} = 4.80\% \times (1 - 0.358) = 3.082\%.$$

Recompute the total take-home yield (still excluding future ROC tax effects):

$$3.082\% + 0.510\% + 0.60\% = 4.192\%.$$

That's how a "great yield" can quietly shrink again. NIIT doesn't show up in the yield quote. It shows up in the gap between what you thought you bought and what you actually keep.

The ``hold" return and the ``sell" return: the two endings investors forget to compare

After-tax returns are tricky because there are two different experiences:

- *The hold experience:* you receive distributions each year, pay taxes each year (in taxable), and keep compounding.
- *The sell experience:* you eventually sell shares and realize a capital gain (or loss). Return of capital makes this especially relevant by reducing basis and potentially increasing that future gain.

If you only calculate after-tax yield and ignore the sale, you're missing the "closing costs" of the investment. If you only calculate after-tax proceeds on sale and ignore years of tax drag, you're missing the slow leak that happened the entire time you owned it.

A professional way to think about after-tax performance is to evaluate both scenarios:

- *After-tax return on distributions during the holding period*
- *After-tax return on price change when you exit*

This mirrors the real world: you live through the years and then you live through the sale. Both matter.

Why a lower yield can beat a higher yield

Once you calculate take-home yield, you start noticing a counterintuitive truth: a lower headline yield in the right account can beat a higher headline yield in the wrong account.

Consider two simplified alternatives:

- *Option A:* A REIT yielding 6% held in taxable, with a take-home yield around 4.2%–4.4% under certain assumptions.

- *Option B:* A more modest 4.5% yield held in a Roth IRA, where the current-year take-home yield can effectively be the full 4.5% (again, depending on the rules and whether withdrawals are qualified).

On the surface, Option A "wins" by yield. In your pocket (or inside the Roth), Option B can win—especially if the Roth dollars are reinvested and compounded without annual tax drag.

It feels unfair until you remember: the Roth is not a different investment. It's a different *container*. And containers change outcomes.

A practical worksheet you can run with your own numbers

You don't need fancy software. You need three inputs and a willingness to be realistic.

1. *Your expected distribution yield* (or trailing yield if that's what you have).
2. *An expected distribution breakdown* (ordinary vs capital gain vs ROC). If you own the REIT already, use last year's 1099-DIV as a starting point.
3. *Your effective tax rates* on ordinary income and long-term capital gains, plus NIIT if applicable.

Then calculate:

$$\textit{Take-home yield (current year)} = Y\left[p_{\text{o}}(1 - t_{\text{o}}) + p_{\text{cg}}(1 - t_{\text{cg}}) + p_{\text{roc}}\right]$$

where:

- Y is the pre-tax distribution yield,
- p_{o}, p_{cg}, and p_{roc} are the proportions of ordinary, capital gain, and return of capital (summing to 1),
- t_{o} is your effective tax rate on ordinary income (including NIIT if applicable),

- t_{cg} is your effective tax rate on long-term capital gains (including NIIT if applicable).

That single formula is enough to stop most yield-chasing behavior in its tracks. It forces the question that matters: "How much of this yield do I actually keep?"

The real win: better decisions, not just better math

The purpose of after-tax calculations isn't to turn you into an amateur tax accountant. It's to keep you from making predictable mistakes with predictable consequences.

When you think in after-tax terms, a few good habits tend to appear naturally:

- You stop treating a REIT's yield as comparable to a qualified-dividend stock's yield.
- You become more careful about putting income-heavy holdings in taxable accounts unless you truly want the income for spending.
- You start valuing "boring" compounding inside tax-advantaged accounts, because you can finally see what the tax drag was stealing from you.

Most importantly, you stop letting the government become your unintended investing partner. In a taxable account, the IRS effectively takes a slice of your REIT income every year, whether or not you needed to pull cash from the portfolio. In a tax-sheltered account, you control the timing (and sometimes the existence) of that tax bill.

That's what after-tax return thinking really does: it hands you back control over a variable most investors leave on autopilot.

If you're going to be picky about anything in REIT investing, be picky about what you keep. The market will handle volatility. Tenants will handle drama. Interest rates will do what they do. Your job is to make sure the return you earn doesn't quietly evaporate on the way to becoming wealth.

Chapter 6

Execution and Lifetime Strategy

You've learned the language of REITs, analyzed the sectors, and built a theoretical portfolio. Now, the rubber meets the road. How do you actually buy without staring at stock tickers all day? When do you sell a winner versus a loser? This chapter is your owner's manual for the long haul-transforming you from a passive observer into a disciplined portfolio manager who spends less time stressing over daily price swings and more time enjoying the passive income stream.

6.1 The Buying Process: Watchlists and Entry Points

Buying a REIT should feel less like "picking a winner" and more like stocking a pantry. You're not trying to predict tomorrow's mood swing in the market; you're trying to assemble a set of durable cash-flow machines you'd be happy to own through boring years, rough years, and the occasional euphoric year when everyone suddenly "discovers" real estate again. The trick is to move the drama out of the moment of purchase. That's what a watchlist is: a way to do your thinking when you're calm, so you can act when the market is not.

Most investors do it backward. They see a headline, a chart, a hot dividend yield, or a friend's excited text, and then they scramble to justify a purchase. The research becomes a courtroom defense: cherry-picked facts, confirmation bias in a suit and tie. A disciplined

REIT buyer does the opposite. The work happens *before* the price looks tempting. When the price *does* look tempting, you're mostly checking that nothing snapped.

A useful mental model: a REIT purchase has two separate questions hiding inside it. First: *Is this a business I would like to own?* Second: *Is this a price I would like to pay?* Watchlists exist to keep those questions from bleeding into each other. If you mingle them, you end up "falling in love" with a low price on a mediocre REIT, or "making peace" with an overpriced REIT because you admire the story. The market is happy to sell you either mistake.

Build the list like a buyer, not a browser

A browser scrolls. A buyer compares. Browsing produces dopamine; buying requires a standard. Your watchlist should behave like a shopping list with rules: if an item doesn't meet the rules, it doesn't get to live on the list, no matter how exciting it looked on television.

Start with a simple truth that feels too obvious to matter: REITs are not all the same product. They rhyme, but they don't replicate. A self-storage REIT, a data center REIT, and a healthcare REIT can all be "real estate" the way a bicycle, a pickup truck, and a forklift are all "vehicles." The steering wheel is not the point. The economic engine is.

So the first filter is a *sector screen*. Not because you're trying to time sectors like a hedge fund, but because sector basics determine what "normal" looks like. Lease length, tenant quality, capital expenditures, sensitivity to recessions, and sensitivity to interest rates vary wildly by property type. A watchlist built without sector awareness tends to become a mirror: it reflects whatever sector happened to be fashionable when you built it.

A practical approach is to pick a handful of sectors you actually understand well enough to explain at a dinner table. Understanding doesn't mean you can forecast the future; it means you know what makes the cash flow grow, what can break it, and what management is allowed to do (and not do) when trouble shows up. If you can't explain the sector's economic drivers in plain English, you're not diversified—you're just scattered.

Once you've narrowed by sector, the second filter is a *quality check*. "Quality" is a fuzzy word, so make it concrete. Quality in REIT land often shows up as: steady cash earnings, sensible leverage, access to capital, and management that behaves like owners rather than fee-collectors. And because REIT accounting has quirks, you usually want to think in terms of cash-flow proxies rather than GAAP net income.

That's where FFO comes in. Funds From Operations (FFO) is a widely used REIT metric that starts with net income and adjusts for items like depreciation that can make real estate companies look artificially unprofitable under standard accounting. Many investors prefer AFFO (Adjusted FFO), which aims to get closer to "cash available for distribution" by subtracting recurring capital expenditures and other adjustments. The key point isn't which acronym wins the argument; it's that you're trying to measure the cash engine, not the accounting noise.

Here's the watchlist-builder's rule: if a REIT can't explain its cash metrics cleanly, reconcile them transparently, and show a history that makes sense, it doesn't belong on the list. Your watchlist isn't a museum of "interesting" tickers. It's a bullpen of candidates you'd actually put on the field.

A funnel, not a rabbit hole

It's easy to turn research into a hobby that never ends. The internet is an infinite buffet of opinions, and your brain is wired to keep sampling. A funnel prevents that. Each step removes candidates for a specific reason, and the list gets smaller on purpose. The goal is not to find *the* best REIT. The goal is to find *enough* great ones that you can wait for price.

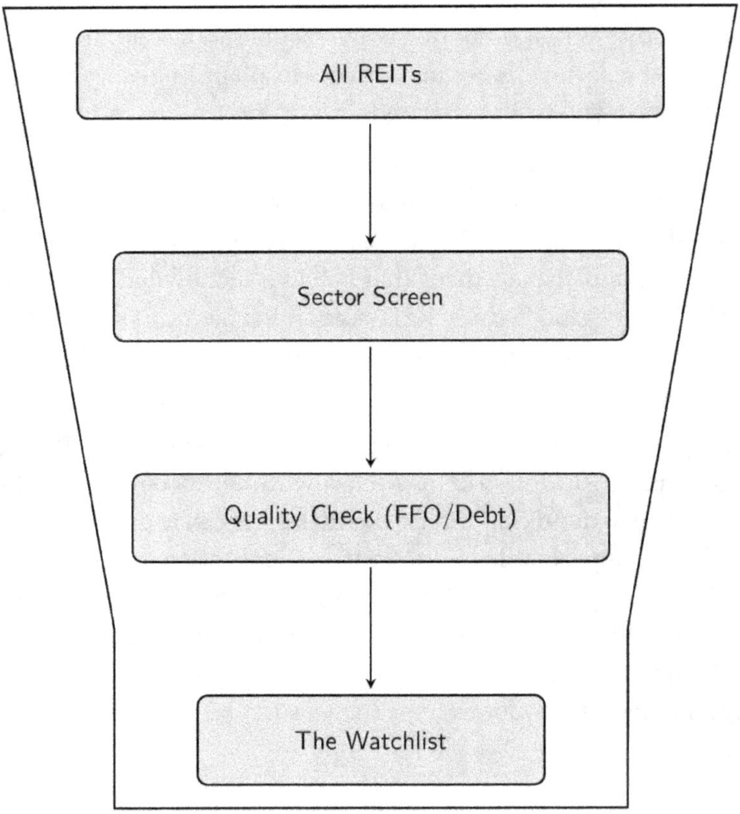

A good funnel step has a pass/fail feel. Not "maybe," not "I kind of like it," but "this meets the bar" or "it doesn't." Your future self—the one staring at a red market day—will thank you for every decision you made in advance.

What ``quality" looks like in plain English

Quality is not perfection. Quality is resilience. It's the ability to keep collecting rent, funding dividends, and refinancing debt without turning into a financial hostage when conditions get tight. In REITs, that resilience tends to come from a few repeatable traits.

First: a business model that doesn't require heroics. If the story depends on everything going right—perfect leasing, uninterrupted rent growth, cheap capital forever—it's not a quality story. A quality REIT can muddle through a mediocre environment and still do fine.

Second: conservative leverage and a mature relationship with debt. REITs use debt; that's not a scandal. The scandal is pretending debt doesn't matter until it suddenly does. When you look at leverage, you're really asking: "How much flexibility does this company have if the world stops cooperating?" A REIT with staggered maturities, ample liquidity, and manageable fixed charges can stay rational in a crisis. A REIT with near-term maturities and thin coverage is one bad quarter away from negotiating with the market on its knees.

Third: a dividend that behaves like a policy, not a dare. Dividends are supposed to be attractive in REITs, but a high yield can be a love letter or a distress flare. The safety question is simple: does the cash flow support the payout with room for maintenance capex, leasing costs, and the occasional unpleasant surprise? Payout ratios—measured against FFO or AFFO—help you quantify that. You're not looking for a magic number; you're looking for a pattern that suggests management isn't trying to win an income beauty contest.

Fourth: management that plays the long game. REITs are capital allocators. They buy, develop, lease, sell, issue equity, refinance debt, and decide whether to grow or defend. A brilliant property portfolio can be undone by sloppy capital allocation. You're looking for a record that suggests management doesn't treat shareholders like a financing source of first resort.

If you want a single sentence to write at the top of your watchlist: *"Would I be comfortable owning this if the stock market closed for five years?"* That question does something wonderful: it makes you care about the business again.

The watchlist is not one list

A single watchlist becomes a junk drawer. A better system is three tiers, each with a different job. This isn't bureaucracy; it's psychological self-defense.

Tier 1: The Bullpen. These are "interesting but unproven" names—companies you're learning, new sectors you're exploring, situations that might become attractive later. The bullpen is allowed to be messy. It's where curiosity lives. You don't buy from the bullpen.

Tier 2: The Bench. These are names that passed your sector and quality screens, but you haven't built enough conviction to act quickly. Maybe you need to read a few more quarters, understand tenant concentration, or see how management behaves in an awkward environment. The bench is a holding pen for "nearly ready."

Tier 3: The Buy List. These are the businesses you'd be happy to own *at the right price.* This list should be short enough that you can name the reasons each belongs there without looking. If your buy list has thirty names, it's not a buy list—it's a directory.

The surprising benefit of tiers is that they reduce impatience. When a new, shiny REIT shows up, your brain wants to skip to ownership. Tiers let you say, "Great, bullpen," and move on with your life. You still get the satisfaction of capturing the idea, but you don't confuse capturing an idea with acting on it.

Valuation: your entry point is a price, not a feeling

A watchlist without entry points is like a grocery list without prices. You can still shop, but you'll overpay whenever you're hungry.

REIT valuation doesn't need to be mystical. You're typically looking at some combination of:

- Price to FFO (or AFFO): what multiple you're paying for the cash earnings.

- Dividend yield: what cash return you're getting at today's price (with the important footnote that yield is only as good as its safety).
- Net Asset Value (NAV) thinking: a rough estimate of what the real estate is worth minus liabilities, compared to the market price.

None of these is perfect. That's fine. The goal is not precision; it's avoiding obvious overpayment.

A practical way to set entry points is to anchor them to *history and plausibility*. What has this REIT typically traded at in normal environments? When it traded cheaper, what was the market afraid of? When it traded richer, what was the market assuming would happen? You're trying to avoid paying "peak optimism" prices for "ordinary" outcomes.

Entry points are best written as ranges, not single numbers. Markets don't hand you the exact price you wrote in your notebook like a polite clerk. Think in zones:

- *Green zone:* clearly attractive valuation relative to history and fundamentals.
- *Yellow zone:* reasonable, but not compelling—good for steady accumulation if you're underinvested or using regular contributions.
- *Red zone:* priced for perfection—only buy here if you have a very specific reason, and "I'm afraid it'll run away without me" is not a reason.

If you want this to work in real life, the zones have to be written down. Otherwise, the zone mysteriously moves the moment the price moves.

The hardest skill: waiting without going numb

Waiting is not passive. Waiting is a decision. And it feels uncomfortable because it's a decision that produces no immediate reward. Buying produces a satisfying "click." Waiting produces nothing but the slow hum of doubt.

The market exploits that discomfort. It dangles motion as a substitute for progress. A REIT rallies and you feel behind. A REIT drops and you feel smart for not buying—but also tempted to buy immediately, because now it's "on sale." Both feelings are traps, just dressed in different outfits.

The antidote is to separate *price movement* from *information.* A price drop is not automatically information. Sometimes it's just a market-wide tantrum. Sometimes it's a real warning. Your watchlist system exists to force the question: "Did the business change, or did the crowd change its mood?" If the business didn't change, a lower price is not a reason to panic. It's a reason to pay attention.

A useful ritual: when a buy-list REIT hits your green zone, do a five-question check before you place an order.

1. Did management change guidance in a meaningful way?
2. Did FFO/AFFO trend break, or is this normal volatility?
3. Did occupancy or same-store NOI weaken beyond what the sector is experiencing?
4. Did the balance sheet get worse (liquidity, maturities, covenants), or is the fear purely macro?
5. Is the dividend still covered, or did the payout ratio quietly become a cliffhanger?

If the answers are stable, you're not "catching a falling knife." You're buying a business you already vetted, at a price you already decided was attractive.

Execution mechanics: how you place the trade matters

Most investors spend hours researching and then place the actual trade like they're ordering fast food. Execution is not a footnote. It's one of the few parts of investing you can control with near certainty.

The basic order types matter, especially in REITs, which can gap on earnings, move sharply around macro news, or trade thinner than mega-cap stocks.

Market orders prioritize getting filled quickly, but not the price you'll receive. In calm markets, the difference may be small. In fast markets, it can be nasty—especially if liquidity is thin. If you hit "buy at market" during a spike, you're essentially telling the market, "Surprise me."

Limit orders flip that tradeoff. You set the maximum price you're willing to pay (for a buy) or the minimum you're willing to accept (for a sell). You get price control, but you might not get filled. That's not a bug; it's the point. A limit order is you acting like a buyer, not a chaser.

A clean REIT practice is to use limit orders for individual names, especially when you're buying because a valuation zone was reached. If you've done the work to decide that $X is attractive and $X + 2% is not, then the order type should reflect that conviction.

A small but meaningful trick: when you place a limit order, don't set it exactly at an obvious round number that thousands of others might also pick. "$50.00" is a magnet. "$49.87" is a decision. You're not trying to be clever; you're trying to reduce the odds that your order becomes part of a crowded price level where fills can be erratic.

Scaling in: a sane compromise between courage and humility

The market has a way of punishing certainty. If you buy all at once, you may feel like a genius or a fool within 48 hours, and neither feeling improves decision-making. Scaling in—buying in multiple tranches—solves a psychological problem more than a mathematical one.

A practical method:

- First tranche when the REIT enters the yellow zone (or when you're building a starter position).
- Second tranche when it hits your green zone.
- Third tranche only if the price gets even better *and* the fundamentals remain intact.

This gives you participation without demanding that you nail the bottom. It also reduces the temptation to "back up the truck" simply because a chart looks dramatic.

Scaling in has another advantage: it forces you to keep checking the business. A lower price is only a gift if the business is the same. Tranches create natural checkpoints where you confirm that you're buying a bargain, not adopting a problem.

Avoiding the two classic entry-point mistakes

REIT investors tend to repeat two mistakes that look opposite but share the same root: letting price action bully the process.

Mistake #1: Chasing yield. A REIT yields 10%, your savings account yields less, and your brain concludes you've found a cheat code. Sometimes a high yield is genuinely an opportunity. Often it's the market pricing in a dividend cut before it happens. The chart looks calm right up until the day it doesn't.

Yield is not the return; yield is a component of return. And in REITs, a dividend cut doesn't just reduce income—it often resets the valuation and damages investor confidence. The question is not "How high is the yield?" The question is "How durable is the cash flow that funds it?"

Mistake #2: Catching the falling knife without looking at the handle. A REIT drops 25% and now it "must" be cheap. But cheap relative to last week is not the same as cheap relative to reality. A REIT can drop 25% and still be expensive if its cash flows are shrinking, its balance sheet is tightening, and its cost of capital is rising.

A disciplined entry process doesn't assume a drop is a bargain. It investigates *why* the market is repricing the asset. Sometimes the market is wrong. Sometimes the market is early. Sometimes the market is late. Your job is to find out which one you're dealing with before you buy.

Write a one-page ``buy discipline" policy

A written policy sounds fussy until you realize it's just a way of borrowing your own judgment from a calmer day. It can fit on one page. It should be blunt enough that you can follow it when you're irritated, tired, or overly excited.

Here's what belongs in it:

- Your buy-list names and the reason each is on the list (one sentence each).
- Your valuation zones (green/yellow/red) and what metrics you use (P/FFO, yield, NAV estimate).
- Your position-sizing rules (starter position size, max position size, and whether you scale in).
- Your default order type (limit orders for individual REITs, with exceptions noted).
- Your "no impulse buys" rule (e.g., any new name must sit in the bullpen for at least one earnings cycle before it can be promoted).

The magic of a policy is that it turns a market day into a checklist instead of a mood. The market can still be noisy; you just stop translating that noise into action.

A final, practical workflow you can actually live with

A watchlist system that requires heroic effort will fail the first time you get busy. The best workflow is the one that feels almost boring.

Once a quarter: refresh the fundamentals for your buy list—FFO/AFFO trends, payout ratios, leverage, debt maturities, and any meaningful changes in tenant or property concentration. Add notes, not essays.

Once a month (or less): glance at valuations relative to your zones. If nothing is in the green zone, do nothing and feel good about it. Doing nothing *on purpose* is a legitimate investing action.

When a name hits green: run the five-question check, place a limit order at your chosen price, and accept that the market may not fill it. Unfilled limit orders are not failures. They're proof you have standards.

That's the heart of disciplined buying: a watchlist that acts like a gatekeeper, entry points that behave like pre-committed decisions, and execution that reflects respect for price. Emotion never disappears—investing would be easier if it did—but a good process keeps emotion from holding the pen.

6.2 The Five-Minute Manager: Effective Monitoring

Owning REITs tempts you into checking your portfolio like it's a sourdough starter: peek, poke, worry, repeat. The problem is that real estate doesn't move at the speed of your phone. A shopping center doesn't re-lease itself because a pundit sounded confident at 2:17 p.m. A warehouse doesn't become obsolete because the ten-year Treasury twitched. Most of what flashes across your screen is *price*, not *progress*. If you monitor a slow-moving business with a fast-twitch habit, you end up managing your emotions, not your investments.

A good monitoring routine has one job: catch meaningful deterioration early enough to act rationally, without training you to flinch at every wiggle. The winning formula is boring on purpose. Quarterly is the natural heartbeat for publicly traded REITs because that's when the real information arrives: earnings, rent trends, occupancy, guidance, balance sheet updates, and the occasional uncomfortable confession. Anything more frequent is usually noise dressed up as urgency.

Think of monitoring like taking vital signs, not doing a full-body MRI every morning. You're not trying to predict next quarter's stock price. You're trying to answer a simpler question: *Is the cash engine still intact?* If yes, you mostly leave it alone. If no, you get curious in a focused way.

The quiet danger of ``staying informed"

There's a psychological trap that feels responsible: constant checking. It masquerades as diligence, but it often produces the opposite. Frequent evaluation magnifies the emotional weight of short-term losses—a pattern behavioral researchers have described as myopic loss aversion. In plain English: when you look more often, you *feel* more risk, even if the underlying long-term risk hasn't changed. The market's day-to-day volatility starts to feel like a verdict on your competence.

REITs are especially good at triggering this because they pay dividends. A dividend arriving can feel like the investment is "working," while a price drop can feel like the investment is "breaking," even if both can happen at the same time for perfectly normal reasons. Monitoring is supposed to protect you from surprises; it should not become a daily audition for your patience.

So here's the framing: your portfolio doesn't need a babysitter. It needs a periodic health check and a clear plan for what would trigger deeper investigation.

A quarterly routine that fits on a sticky note

The simplest effective cadence is built around four moments each year: when your REITs report results. Pick a small window—say, one weekend after most of your holdings have reported—and do the same quick scan every time. If you want this to last decades, it has to be light enough that you'll still do it during busy seasons of life.

A practical quarterly check can be five minutes per holding once you know what you're looking for. The trick is to standardize the questions so you're not reinventing your process every time.

What follows is a set of signals worth tracking, plus the noise you can safely ignore most of the time.

Signal #1: FFO (or AFFO) trend---the engine, not the paint

Funds From Operations (FFO) is a REIT earnings measure meant to approximate the cash-generating power of the properties by adjusting GAAP net income for items like depreciation. Many REITs also report AFFO (Adjusted FFO), which tries to get closer to "cash left over" after recurring costs. The important point is not memorizing the acronym; it's watching whether the cash engine is getting stronger, stable, or weaker.

Quarter to quarter, FFO can bounce for harmless reasons: timing of acquisitions, lease-up costs, one-time fees, straight-line rent accounting, storms, or property sales. The pattern matters more than the tick. Ask two questions:

- *Is the trend broadly intact over the last year or two?*

- *Does management's explanation make economic sense, or does it feel like storytime?*

You're looking for deterioration that persists. A single soft quarter can be weather. Two or three quarters of slippage with vague explanations is often a business issue wearing a weather costume.

One subtle but crucial habit: when a REIT reports "FFO," glance at the reconciliation. FFO is widely used, but it is not a single, perfectly standardized number across every company. If a management team starts adding back more and more "non-recurring" items every quarter, that's not an adjustment—that's a personality trait.

Signal #2: Same-store NOI and occupancy---the properties speaking for themselves

Net Operating Income (NOI) is essentially property-level profit before corporate overhead and financing. Same-store NOI tries to measure the performance of properties owned in both periods, filtering out the noise of acquisitions and dispositions. It's one of the cleanest ways to hear what the buildings are actually doing.

Occupancy is the simpler cousin: how full are the properties? But don't stop at the percentage. Occupancy can be held up cosmetically with aggressive leasing that sacrifices rent rates or tenant quality. When you read an earnings release, try to connect three dots:

- occupancy change,
- rent growth (often discussed as "same-store rent" or leasing spreads),
- and same-store NOI growth.

When those three start telling conflicting stories, it's worth slowing down. A REIT that is "full" but experiencing weak rent growth may be facing tenant bargaining power or oversupply. A REIT with strong rent growth but falling occupancy could be pushing rents too hard and risking a vacancy spiral. Neither is automatically fatal; both are worth noticing early.

If you want one memorable line to keep your eyes honest: *occupancy is the pulse, but rent is the oxygen.*

Signal #3: Dividend coverage---treat the dividend like a budget item

Dividends are the emotional center of REIT investing. They're also where denial likes to hide. Investors often assume the dividend is safe because it *has been* safe, or because the company has a long history of paying. History matters, but cash flow matters more.

Dividend coverage can be approximated by a payout ratio: dividends divided by FFO or AFFO. A lower payout ratio generally means more cushion. A higher ratio can still be fine in stable sectors with predictable cash flows, but it leaves less room for surprises, refinancing costs, and necessary property spending.

Two monitoring moves help keep you grounded:

- *Watch the payout ratio trend, not just the current number.* A slow creep upward can be the first sign that the dividend is being maintained by hope.
- *Read the dividend declaration like a signal.* Is it steady? Growing modestly? Frozen? Cut? A sudden freeze after a period of increases can be management's quiet way of saying, "We see headwinds."

The dividend is not a trophy. It's a policy choice funded by cash. When cash gets tight, the dividend becomes negotiable whether investors like it or not.

Signal #4: The balance sheet---debt maturity schedules and liquidity

REITs live in the real world, which means they borrow money. The danger is rarely "having debt." The danger is *having debt at the wrong time.* Monitoring debt is mostly about timing and flexibility.

Your quarterly scan should answer four simple questions:

- *How much liquidity does the REIT have?* (Cash on hand plus availability on credit lines.)
- *When does the next meaningful chunk of debt mature?*
- *Is the debt mostly fixed-rate or floating-rate?*
- *Is leverage stable, rising, or falling?*

You don't need to become a credit analyst. You're looking for a profile that allows management to refinance from a position of strength. A REIT with ample liquidity and staggered maturities can wait out ugly credit markets. A REIT with a wall of maturities next year is at the mercy of whatever interest rates and lending standards happen to exist at that moment.

One especially useful mindset: refinancing is not a philosophical event; it's a math event. Higher rates raise interest expense. Higher interest expense squeezes FFO. Squeezed FFO pressures the dividend and reduces growth capital. That chain reaction is why the debt schedule belongs on your "signal" list even if you find debt tables about as thrilling as appliance manuals.

Signal #5: Guidance and management credibility

REIT management teams give guidance because investors demand it, but guidance is more than a number. It's a window into how management thinks.

When guidance changes, don't only ask, "Up or down?" Ask, "Why?" A high-quality explanation sounds like operations: leasing conditions, tenant behavior, expense pressures, development timelines, and financing costs. A low-quality explanation sounds like atmosphere: "macro uncertainty," "fluid environment," "transitory headwinds," and other phrases that translate to "please stop asking."

You're also monitoring for credibility drift. If a management team consistently misses guidance, redefines "non-recurring," or promises that every challenge is temporary, your risk isn't just the properties. Your risk is the decision-making.

A pleasant surprise is nice. A nasty surprise is expensive. Credibility reduces the odds of nasty surprises.

The ``what to ignore" list (most of the time)

Monitoring works best when it includes deliberate blindness. Not ignorance—selective attention. The market sells you an all-you-can-eat buffet of information; your job is to decide what is nutritious.

Here are common noise sources that deserve a low-priority seat:

- *Daily price moves.* They are often the market reacting to itself.
- *Generic macro headlines.* "Rates up" and "rates down" are not company-specific analysis.
- *Pundit commentary.* Many opinions are produced to fill airtime, not to improve your returns.
- *One-off news items that don't change property cash flow.* A scary-sounding article that doesn't affect occupancy, rent, expense structure, or financing is usually entertainment.

This doesn't mean you never react to news. It means you react *through the lens of your signals.* If a headline matters, you should be able to translate it into one of your monitored metrics: FFO, NOI, occupancy, dividend coverage, or debt/refinancing risk. If you can't translate it, it probably isn't actionable.

A one-page dashboard you can reuse forever

If you want to turn five minutes into something repeatable, use a simple template for each holding. Keep it consistent. Consistency is what makes the process fast.

A clean dashboard can live in a spreadsheet, a note-taking app, or a printed page in a folder. It doesn't need to be fancy. The power is that each quarter you update the same fields and the trend becomes visible without effort.

Here's an example layout using LaTeX and `tabularx` that you can adapt:

REIT Name / Ticker	
Property Type / Thesis in one line	(Example: "Industrial warehouses focused on infill markets; demand driven by logistics and constrained land.")
FFO or AFFO trend (YoY)	(Up / Flat / Down; short note on why)
Same-store NOI / Occupancy	(Key numbers; note leasing spreads if provided)
Dividend	(Current rate; payout ratio estimate; recent change?)
Balance sheet	(Liquidity; next big maturity; fixed vs floating; leverage trend)
Guidance	(Raised / maintained / cut; reason; confidence level)
Red flags / Watch items	(Anything that would trigger deeper research)

Notice what's missing: today's stock price, a chart, and whatever social media is arguing about. Those aren't vital signs. They're mood rings.

When to escalate from ``five minutes" to ``deep dive"

A light routine works because you don't deep-dive every quarter on every holding. Deep dives are for when the signals change.

Escalation triggers tend to fall into a few categories:

- *A sustained FFO/AFFO decline* that isn't explained by a temporary, well-defined cause.
- *A payout ratio that jumps* because cash flow fell or the dividend rose without support.
- *A sudden occupancy drop* or weakening same-store NOI that suggests demand is cracking.
- *A tightening debt situation* (near-term maturities, shrinking liquidity, rising floating-rate exposure, covenant risk).
- *Management credibility issues* (surprise announcements, repeated "adjusted" optimism, unexplained misses).

Escalation doesn't automatically mean "sell." It means you stop treating the holding as a background asset and start treating it as an active question. The goal is to diagnose early, while you still have options and while management is still speaking in full sentences rather than emergency press releases.

The real output of monitoring: calmer decisions

Monitoring feels like it should produce action. In practice, its best output is restraint. When the market drops and your screen turns red, your brain wants certainty. A good quarterly routine gives you something better: evidence.

Evidence lets you say, "The price is down, but occupancy is stable, same-store NOI is fine, the dividend is covered, and maturities are manageable." That sentence is worth money, because it keeps you from turning a temporary markdown into a permanent mistake.

And sometimes evidence tells a different story: "FFO is slipping, the payout ratio is climbing, and debt maturities are getting close." That sentence is also worth money, because it prompts you to act while the situation is still manageable, not after the market has already finished repricing the risk.

The five-minute manager isn't lazy. The five-minute manager is *selective*. Real estate rewards selectivity. It's a slow asset class, built on leases, locations, and financing. Your monitoring should match that tempo: calm enough to be consistent, sharp enough to be useful, and disciplined enough to ignore the daily circus outside the properties.

6.3 Knowing When to Fold: Selling and Rebalancing

Selling is where investors accidentally turn a sensible REIT plan into performance art. Buying has guardrails—watchlists, entry points, limit orders, patience. Selling, by contrast, often happens in a hurry, with a knot in your stomach and a CNBC chyron screaming in the background. The market knows this about us. It offers fear at wholesale and buys your conviction at retail.

A clean selling framework does something surprisingly comforting: it makes most selling decisions *unnecessary*. Not because nothing ever goes wrong, but because you stop treating price drops as emergencies. A REIT can be down 10% for a dozen reasons that have nothing to do with its properties. The question is not "Why is the stock down?" The question is "Did the business change in a way that permanently hurts the cash engine?"

That word *permanently* is the hinge. Real estate is full of temporary discomforts: a tenant moves out, a development gets delayed, interest rates rise, a recession bruises demand. The good operators absorb those hits, adjust, and keep collecting rent. What you're hunting for is different: structural damage. The kind that changes what "normal" will look like from here on.

If you want a blunt rule that saves an enormous amount of stress, it's this: *there are only three defensible reasons to sell a REIT*. Everything else is mood management.

- *Thesis break:* the business is permanently impaired.
- *Extreme overvaluation:* the price becomes detached from reality.
- *Rebalancing:* the position size becomes a risk problem, even if the REIT is fine.

Notice what's missing: "the price is down," "the price is up," "I'm bored," and "someone on the internet hates it." Those are feelings, not reasons.

The first rule of selling: diagnose before you treat

Most people sell for the same reason they slam the brakes in a parking lot: they saw movement and assumed danger. But in investing, movement is normal. A REIT can drop because rates moved, because a sector fell out of fashion, because a big ETF rebalanced, or because the market is throwing a tantrum about something adjacent. Price is a smoke alarm that goes off when you make toast. Sometimes it's a fire; sometimes it's breakfast.

So the first step is always diagnosis. When a REIT drops, your process should force three questions in order:

- Has the cash engine (FFO/AFFO) actually weakened?
- Is the dividend still covered with a reasonable cushion?
- Has the original thesis broken, or is this a temporary headwind?

That order matters. If you start with "thesis," you can rationalize anything. If you start with cash and coverage, you anchor yourself in reality.

Right in the middle of that diagnostic mindset sits the most useful mental reframe in REIT investing: *you are not managing a stock chart; you are managing a claim on rent checks.* Stock charts are loud. Rent checks are quiet. Quiet is usually where the truth lives.

The thesis break: what it looks like when the floor actually gives way

A thesis break is not "things got harder." It's "the rules changed." It's when the economic engine you thought you owned turns into a different engine—one that produces less cash, more volatility, and fewer good options.

In REITs, thesis breaks often come from a handful of culprits:

Demand got structurally weaker. Not "a recession slowed leasing," but "this property type is being used less, permanently." The classic example is certain kinds of retail that never recovered foot traffic because consumer behavior rewired. Another example is an office portfolio that loses its pricing power because tenants downsize and renew on tougher terms for years, not quarters.

The balance sheet becomes a trap. REITs can survive an ugly operating year. What kills them is being forced to refinance or raise capital at the worst possible moment. A wall of near-term maturities, shrinking liquidity, and rising interest expense can turn an operating problem into an existential one. When the cost of capital jumps, growth plans get canceled, asset sales become desperate, and shareholders get diluted at low prices.

The dividend becomes fiction. A dividend cut is not automatically a thesis break. Sometimes it's a responsible reset. But a pattern of "supporting" the dividend with asset sales, aggressive add-backs in AFFO, or optimistic guidance that never arrives is the financial equivalent of paying rent with a credit card. It works until it doesn't.

Management behavior changes. This one is underrated. If a management team starts issuing equity at cheap prices to fund empire-building, or repeatedly "adjusts" away recurring expenses, or makes acquisitions that don't fit the strategy, your thesis may break even if the properties look fine. Real estate is forgiving; capital allocation is not.

A thesis break demands honesty, and honesty is expensive because it forces you to admit you were wrong. Investors don't avoid thesis breaks because they're rare. They avoid them because they bruise the ego. That's why the decision tree below is useful: it removes ego from the driver's seat.

Decision tree: when a price drop is a gift, and when it's a warning

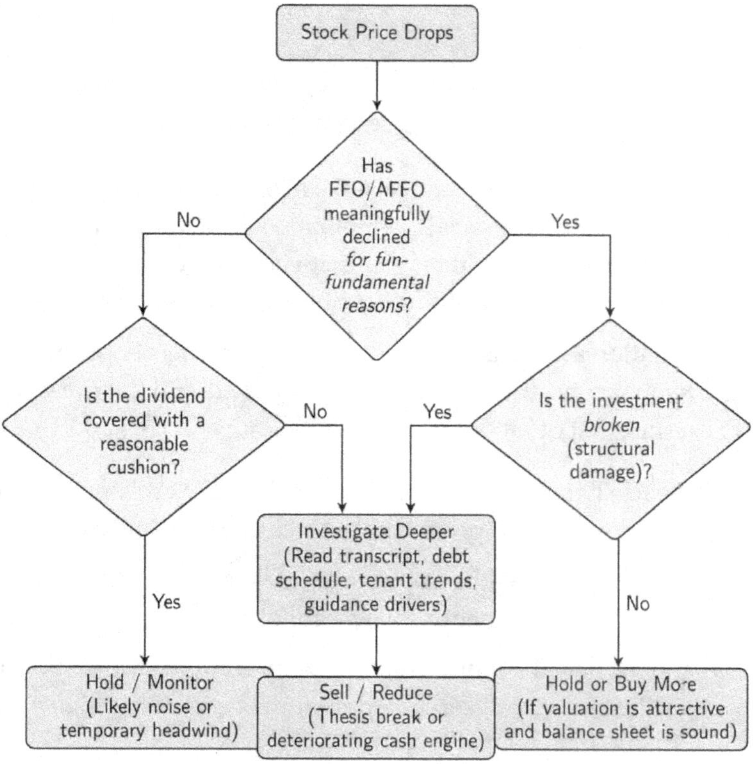

The power move here is that "Investigate Deeper" is an actual box. Most investors skip it. They jump from "price dropped" to "sell" or "buy more" like they're flipping a coin. Investigation is the bridge between a chart and a business. It's where you look for whether the decline is operational (leases, occupancy, rent spreads), financial (interest expense, maturities, liquidity), or simply market sentiment.

And notice the asymmetry: a price drop with stable fundamentals is allowed to be a buying opportunity, but a price drop with a weakening cash engine is treated like a warning until proven otherwise.

That's how you avoid catching falling knives without turning into a chronic do-nothing.

Extreme overvaluation: selling because the market got silly

Selling due to overvaluation feels like showing up early to a party and leaving before dessert. It's emotionally difficult because prices can stay irrational longer than your patience, and because greed has a very persuasive voice: "What if it doubles from here?"

Overvaluation selling is not about being clever. It's about recognizing when your future returns have been pulled into the present. In REIT terms, that often shows up when:

- the REIT trades at a multiple of FFO/AFFO far above its own history *and* above peers without a clear, durable reason,
- the dividend yield compresses to a level that implies heroic growth or permanently cheap capital,
- and the narrative shifts from property economics to stock-price justification ("it deserves a premium because everyone wants it").

Here's the key: overvaluation is rarely a *precise* call. That's why "sell all" is often the wrong reflex. A more realistic move is trimming. If you love the business but hate the price, reducing the position is a way to respect both truths at once.

A disciplined overvaluation trim also keeps you from committing a common REIT sin: letting one spectacular winner quietly become your entire portfolio. Concentration often happens not because you chose it, but because you didn't notice it.

Rebalancing: the unglamorous skill that keeps you solvent

Rebalancing is the grown-up version of "don't put all your eggs in one basket," except it's actually enforceable. It answers a question most investors avoid: *How much of this one bet am I willing to carry?*

A simple rebalancing policy can be calendar-based (check weights every 6 or 12 months) or threshold-based (rebalance when an allocation drifts beyond a preset band). The specific method matters less than the fact that you have one. Without a policy, you rebalance when you feel like it, which usually means you rebalance *after* the market has already made it painful.

A threshold policy is especially good for REIT investors because REITs can move in clusters. If one sector runs hot, your portfolio can drift from "diversified" to "accidentally thematic" without you noticing. Thresholds are an alarm that goes off before your risk becomes a headline.

Here's a practical way to set bands without turning your portfolio into a spreadsheet competition:

- Set a *maximum position size* for any single REIT (for many investors, something like 3%–7% depending on portfolio size and risk tolerance).
- Set a *sector cap* (so one property type can't dominate your outcomes).
- Decide on a *rebalance trigger*: either a date on the calendar or a drift amount (for example, a position exceeds its cap by 25% relative, or 2 percentage points absolute).

Rebalancing is not a comment on whether a REIT is "good" or "bad." It's a comment on how much damage you're willing to tolerate if you're wrong about anything—rates, tenants, management, regulation, or plain bad luck.

The tax reality: REIT distributions are not all the same animal

Selling and rebalancing decisions live in the messy world of after-tax outcomes, and REITs add a twist: distributions can arrive in different forms. In a taxable account, REIT payouts may include ordinary income, capital gain distributions, or nondividend distributions (often return of capital that reduces your cost basis). That means two REITs with the same headline "yield" can produce different tax results, and trimming a position can interact with your cost basis in ways that surprise you at tax time.

This is not a reason to avoid rebalancing. It's a reason to do it with eyes open. If you're holding REITs in both taxable and retirement accounts, you may have flexibility: rebalance inside tax-advantaged accounts to reduce taxable events, then use new contributions in taxable accounts to nudge weights back toward target. The cleanest rebalancing is often the kind you do with cash flows rather than sales.

A useful habit: when you consider selling in a taxable account, pause and ask, "Am I selling because the business changed, or because my allocation changed?" Those are different problems, and they deserve different solutions.

How to sell without giving away money: execution still matters

Once you've made a rational sell decision, the final insult would be to execute it carelessly. Just as buying with a market order can produce an unpleasant surprise, selling with the wrong order type can turn a controlled decision into a sloppy one.

For most liquid, widely traded REITs, a limit order can be a sensible default, especially in volatile markets. It prevents you from selling into a temporary air pocket. The tradeoff is that a limit order might not execute if the price falls quickly. That's not always bad. If your thesis is truly broken and you need out *now*, immediacy may matter more than price control. But when you're trimming for rebalancing, patience is often rewarded.

Stop orders and stop-limit orders can be useful tools, but they come with sharp edges. A stop order becomes a market order when triggered, which means the fill price can be worse than expected in a fast move. A stop-limit order preserves price control, but it can fail to execute if the price gaps past your limit. The point isn't to ban these tools; it's to treat them like power tools. Great in skilled hands, ugly when used casually.

If you're going to automate anything, automate your *rules*, not your panic. Predetermine the conditions under which you would reduce a position, and the order type you'll use, when you're calm.

A realistic "sell file" to keep you honest

Investors keep buy lists. Almost nobody keeps a sell file. That's a mistake, because memory is a biased narrator. A sell file is simply a short note for each holding that answers:

- What would have to happen for this to be a thesis break?
- What valuation would make me consider trimming even if the business is fine?
- What position size would force rebalancing regardless of my enthusiasm?

The point is not to predict the future. The point is to reduce the odds that you improvise under stress. If you've ever sold a perfectly good REIT at the bottom because you couldn't stand watching it fall, you already understand why this works.

The calm ending most investors never experience

The best selling decisions rarely feel heroic. They feel almost clinical. You reviewed the cash engine, dividend coverage, balance sheet, and thesis. You decided the business was fine and held—or even added at a better price. Or you decided something fundamental cracked and you reduced or exited before the crack became a collapse. Or you trimmed because a winner grew too large and you refused to let success turn into fragility.

That's what it looks like when a portfolio is managed rather than endured: fewer dramatic moments, fewer regret trades, and more outcomes driven by property economics instead of market theater. When you get selling and rebalancing right, you don't feel like a genius. You feel like someone who sleeps normally.

6.4 Retirement Strategy and Global Opportunities

Retirement has a strange psychological twist: the moment your portfolio is finally big enough to matter, every decision feels like it carries more weight. In your working years, a bad month is annoying. In retirement, a bad month can feel like a threat to the roof over your head. That emotional pressure pushes people into two equally expensive mistakes: hoarding cash out of fear (guaranteeing inflation will quietly pick your pocket) or chasing yield out of urgency (guaranteeing the market will eventually hand you a lesson with interest).

REITs can be a stabilizer here, but only if you treat them as what they really are: a liquid claim on rent-producing property portfolios, packaged inside a stock wrapper. The stock wrapper is noisy; the rent is not. A good retirement strategy leans into the rent-like quality without forgetting the wrapper can still swing around like a screen door in a storm.

The goal isn't to design a plan that never feels uncomfortable. The goal is to build a plan that still works when it *does* feel uncomfortable—because it will.

Two retirement mindsets: living off yield vs. living off the portfolio

Most retirees naturally gravitate to the idea of "living off dividends." It feels clean. Spend the income, keep the principal, sleep well. And REITs, with their historically higher distribution rates compared to many broad equity indexes, make that story feel especially tempting.

There's nothing wrong with the *natural yield* approach (spending dividends and interest), as long as you understand its hidden trade: you're outsourcing your spending policy to whatever your holdings decide to distribute. If a REIT cuts its dividend in a rough patch, your "paycheck" drops. The portfolio might be fine long-term, but your month-to-month budget suddenly isn't.

The alternative is the *total return* approach: you set a sustainable spending target and fund it from a mix of dividends *and* occasional sales. That can sound scary, mostly because selling shares feels like consuming the machine instead of consuming the output. But in practice, total return can be more stable because you're not hostage to dividend policy. You're choosing a withdrawal rate and using the portfolio's growth and income together.

Here's the useful reframing: selling shares is not automatically "breaking" the portfolio. A portfolio is a tool for converting capital into future spending. The real risk is not sales; it's *overspending relative to what the portfolio can support*.

REITs fit into both mindsets. They can provide meaningful cash flow so you sell less in down markets, and they can also be part of a diversified total return plan where you occasionally trim what's grown and add to what's cheap.

A practical hybrid: dividends as the baseline, total return as the backstop

In real retirement life, the cleanest strategy is often a hybrid. Dividends fund the predictable, boring bills: groceries, utilities, insurance, property taxes. The portfolio's flexibility—a small planned sale here and there—funds the lumpy stuff: a roof, a car, travel, helping family, the "surprise" medical bill that arrives like it owns the place.

This hybrid approach has a behavioral advantage: it keeps you from obsessing over whether you "had to sell." If the market is down, dividends still arrive, and you can choose to delay discretionary spending or pull from a cash buffer instead of selling at an ugly price. If the market is up, trimming a bit to refill cash feels like harvesting, not sacrifice.

REIT dividends are especially suited to this baseline role because they tend to be meaningful in size. That's not a guarantee of safety—you still care about payout ratios and balance sheets—but it does mean a thoughtfully built REIT allocation can reduce the frequency with which you have to sell into volatility.

Sequence-of-returns risk: the retirement villain that doesn't announce itself

The market has a nasty habit of delivering its worst declines right after someone retires, when the portfolio is most vulnerable. This is sequence-of-returns risk: poor returns early in retirement can cause disproportionate damage because withdrawals lock in losses by shrinking the base that would otherwise recover.

REITs don't magically fix sequence risk. Public REITs can drop hard during equity selloffs and rate shocks. What they *can* do, when chosen carefully, is provide a stream of cash flow that reduces the need to sell at the worst moments. Think of dividend income as friction in the system: it slows down the destructive compounding of forced sales.

A retiree-friendly implementation is simple in concept:

- Maintain a modest cash buffer (enough to cover a chunk of spending so you're not selling reactively).
- Let dividends replenish that buffer in normal markets.

- Refill the buffer with planned trims during strong markets.

This is less about maximizing returns and more about minimizing panic.

Required Minimum Distributions: the government's withdrawal schedule

Even if your personal spending plan is elegant, the tax code may insist on its own timetable. Required Minimum Distributions (RMDs) are the IRS rule that says certain retirement accounts can't be left untouched forever. If you have traditional IRAs or many workplace plans, you generally must begin taking minimum withdrawals when you reach a certain age. The current general starting age is 73.

The RMD mechanics are not complicated, but they're unforgiving in the way airport security is unforgiving: the line is simple, the consequences of missing it are unpleasant. The first RMD has a special deadline—it can generally be taken as late as April 1 of the year after you reach the required age—but delaying that first withdrawal can create a tax headache because it may force *two* taxable distributions into the same calendar year.

After that initial grace period, the rhythm is straightforward: RMDs are generally due by December 31 each year.

There's also a penalty regime that makes procrastination a particularly bad hobby. If you fail to take an RMD (or take too little), the IRS can assess an excise tax on the shortfall. The baseline rate is 25%, and it can be reduced to 10% if corrected in a timely manner.

None of this is meant to scare you; it's meant to keep you from accidentally turning a paperwork mistake into an expensive one. The operational solution is boring and effective: consolidate old accounts where appropriate, keep a simple annual checklist, and automate what you can.

What RMDs mean for a REIT investor (and why it's not automatically bad)

RMDs are often described as a "problem," but they're really just forced distribution of taxable income that was deferred for decades. If you already planned to live on your portfolio, RMDs may simply replace withdrawals you would have made anyway.

The more interesting question is *where the RMD cash comes from*. Many retirees let the custodian sell whatever is easiest, which is how you end up liquidating a high-quality REIT position at a bad time because it happened to be the most liquid thing in the account.

A more intentional approach treats RMDs as part of rebalancing:

- If REITs have run up and your allocation is above target, selling REIT shares to satisfy part of the RMD can be an elegant trim.
- If REITs are down and you believe the fundamentals remain intact, you might satisfy the RMD by trimming something else (or using cash) and allow the REIT position time to recover.

The point is not to "game" the market. It's to avoid accidental, automatic decisions that fight your own portfolio discipline.

The tax personality of REIT income: not all ``dividends" behave the same

REIT distributions are often lumped into the single word "dividend," but the tax reporting can be more nuanced. In taxable accounts, distributions can include ordinary dividends, capital gain distributions, and nondividend distributions (often return of capital that reduces your cost basis). If you're planning retirement cash flow, those distinctions matter because they affect how much you owe and when.

Even if you never become a tax nerd, there's one document that deserves a small slice of your attention every year: Form 1099-DIV. It's where the story of your distributions gets translated into tax categories.

Two boxes deserve special mention for REIT investors:

- *Box 5* (*Section 199A dividends*): this reports qualified REIT dividends that may be eligible for the Section 199A deduction.
- *Box 7* (*Foreign tax paid*): this becomes relevant if you own international funds or REITs that have foreign withholding.

There's also a holding-period rule wrapped into "qualified REIT dividends" for Section 199A purposes: the benefit can depend on whether shares were held long enough around the ex-dividend date. This is a quiet reason retirement investors often do better with steady ownership rather than frequent trading: the tax code is friendlier to calm behavior.

Section 199A after 2025: the rule that reminds you politics exists

For years, many investors treated the Section 199A deduction as a helpful tailwind for REIT dividends, but with an asterisk: it was originally scheduled to sunset after 2025 under the TCJA framework. In lifetime planning, sunsets matter. A strategy that depends on a tax break is less of a strategy and more of a wish.

As of IRS guidance published in the Internal Revenue Bulletin, Section 199A has been amended for taxable years beginning after December 31, 2025 (including the creation of a minimum deduction and related thresholds), which strongly implies continuation rather than a simple expiration.

The right attitude is pragmatic: enjoy tax benefits when they exist, but don't build a retirement plan that collapses if Congress changes its mind later. Build a spending policy that works without heroic assumptions, and treat tax rules as variables you monitor annually.

International REITs: diversification with an accent (and a few extra receipts)

At some point, a thoughtful REIT investor asks a reasonable question: "If real estate is local, why am I only investing in my local market?" It's a fair instinct. The world has shopping centers, cell towers, apartments, warehouses, and data centers everywhere. Owning a slice of global property cash flows sounds like common sense.

International REIT exposure is often accessed through index-based products. Two widely referenced benchmark families for listed real estate are FTSE EPRA Nareit and MSCI.

Those benchmarks are helpful not because an index is "safe," but because it forces a definition: what counts as listed real estate, what countries are included, how constituents are selected, and how the portfolio is maintained. If you buy a fund that says "global real estate," the benchmark tells you what that actually means in practice.

What global exposure can genuinely improve

Global REITs can offer diversification across:

- *Economic regimes:* different countries experience different growth, inflation, and property cycles.
- *Sector composition:* some regions lean more heavily into certain property types.
- *Policy environments:* interest-rate paths and regulatory choices diverge across central banks and governments.

The best-case outcome is not that international REITs outperform domestic REITs every year. The best-case outcome is that they zig when your home market zags, smoothing the ride and giving you more options for rebalancing.

But there's a blunt truth worth saying out loud: diversification benefits can shrink during global stress. In a crisis, correlations often rise because investors sell what they can, not what they should. Global exposure is a seatbelt, not a force field.

The three extra risks you inherit when you go global

International REITs come with additional moving parts. None are fatal; all deserve respect.

Currency risk. If you buy a Japanese REIT and the yen weakens against the dollar, your U.S.-dollar returns can suffer even if the REIT does perfectly fine in local terms. Currency can amplify gains or erase them, and it has a talent for moving for reasons unrelated to property fundamentals.

Withholding taxes and tax paperwork. Many countries withhold tax on dividends paid to foreign investors. In a taxable account, you may be able to claim a foreign tax credit for eligible withholding. This often shows up via foreign taxes paid reported on tax documents, and the correct treatment can depend on treaty rates and your personal situation.

Different rules of the road. REIT regimes and corporate governance standards vary. Some markets have structures that look like U.S. REITs; others have listed property companies with different payout norms, leverage profiles, and regulatory constraints. Index providers try to standardize exposure, but they can't standardize legal systems.

The practical message is simple: global REITs can be worth owning, but they are not "just U.S. REITs with different zip codes."

A sensible way to add global REITs without turning your portfolio into a geography project

If you decide global exposure is worth it, simplicity is usually your friend. Many investors are best served by using a broad, low-cost fund tied to a well-known benchmark rather than assembling a patchwork of single-country positions. Benchmarks such as the

FTSE EPRA Nareit index family and MSCI's developed-market real estate indexes are common reference points in how funds define "global listed real estate."

A second sensible rule: size the allocation so it helps diversification without hijacking your attention. If you add a small sleeve of international REITs, you're buying optionality and smoothing potential. If you make it huge, you're also signing up to care deeply about currency moves, foreign policy, and tax withholding nuances. That can be interesting. It can also be exhausting.

And exhaustion is not a portfolio strategy.

Retirement is an operations game as much as an investing game

In accumulation years, investing can feel like picking. In retirement, investing becomes scheduling: scheduling withdrawals, scheduling tax payments, scheduling rebalancing, scheduling RMD compliance, and scheduling your own sanity so you don't spend your best years staring at a brokerage app.

REITs can play a starring role in that operations game because they produce cash flow and because they're liquid. Liquidity is underrated until you need it. When the furnace dies or a family member needs help, you don't want to be trapped in a position that can't be sold or that forces you to accept a terrible price.

A retirement strategy that actually holds up has three features:

- a spending policy that doesn't depend on perfect markets,
- a tax routine that respects deadlines and avoids surprises,
- and a portfolio design that gives you choices when the world gets inconvenient.

That's the quiet promise of doing this well: not maximizing every last basis point, but maximizing the number of years you get to think about your life instead of your account balance.

www.ingramcontent.com/pod-product-compliance
Lightning Source LLC
LaVergne TN
LVHW022207200926
848481LV00006B/1882
9798896653134